D0097719

The High Road

Related Books by Ben Bova

FICTION

Millennium
The Multiple Man
Colony
Kinsman
Voyagers

NONFICTION

In Quest of Quasars
The Fourth State of Matter
The New Astronomies
Starflight and Other Improbabilities
Closeup: New Worlds

The High Road

BEN BOVA

Houghton Mifflin Company Boston 1981

Library of Congress Cataloging in Publication Data

Bova, Ben, date
  The High road.

  Includes index.
    1. Outer space—Exploration.   2. Astronautics and
civilization.   I. Title.
TL793.B677   338.9'00999   81-6304
ISBN 0-395-31288-4   AACR2

Printed in the United States of America

S 10 9 8 7 6 5 4 3 2 1

Portions of this book have appeared previously in *Omni*
and *Penthouse*.

To Barbara,
    with all my love and gratitude

Preface

THE "HIGH ROAD" OF SPACE is where Ben Bova wants us to travel. It is where I have traveled and would have my country travel again. It is where millions of Americans see their future and the future of their children.

The high road of space has been mapped and paved by the energy and dedication of more than 500,000 young Americans. It offers no technological barriers, only technological advances for humankind. It offers no inherent danger, only freedom for those who choose to travel. It offers no stifling of the human spirit, only an infinitely variable expansion of that spirit to and beyond the bounds of the solar system.

The high road of space has no inherent potential for nationalistic or totalitarian dominance. It does offer the means to extend freedom to other places and other times and to protect our civilization here on Earth.

Ben Bova's *The High Road* provides a unique vision of the past, present, and future. With that vision, he gives us the ultimate challenge: Do we care enough about the survival of the human race, and the survival of our society, to put our children on the high road of space?

You must search for your answer to this challenge in the pages of this book. I found my answer, an unequivocal *yes*, in

Norway in 1957 as a young student among students from all over the world. The flight of Sputnik I overhead had such a profound emotional and political effect on the view of the future by these young people that I vowed to be involved in some way.

First of all, I was angry and not a little scared because that beeping image of the future had a red star on it. On the other hand, the thoughts about what could happen next were exciting. John Kennedy's statement of purpose for the nation in space became a statement of purpose for a generation. For us and the nation, the Apollo program gave a purity of purpose that provided an outlet for all the emotions of accomplishment that other generations had known. For us, it overshadowed the international and civil strife of the sixties, symptoms of man's inhumanity, with a tangible expression of the inherent freedom and creativity of humanity.

Apollo showed us all we need to know about what is possible in space. It also confirmed that the future history of civilization will be determined by what happens in the new ocean of space, just as the past history of civilization was determined by what happened on the oceans of Earth.

But we do not have much time. A new and permanent purpose in space must envelop this nation if we are to survive the twentieth century and prosper in the twenty-first. We have no other acceptable choice but to move forward.

The High Road, under Ben Bova's skillful hand, will guide you through the accomplishments and mistakes of the past, the logic of the present, and a startling view of the future. Enjoy this trip, but remember that the parents of the first Martians playing outside in your yard expect more from you than that.

HARRISON H. SCHMITT
United States Senator
and Apollo 17 Astronaut
Silver City, New Mexico
March 6, 1981

CONTENTS

x   *Contents*

I

THE PROBLEM

These are the times that try men's souls. The summer soldier and the sunshine patriot will, in this crisis, shrink from the service of his country . . .

—THOMAS PAINE

1

The Real Space Race

Let's talk sense to the American people.
Let's tell them the truth, that there are no
gains without pains.
—ADLAI E. STEVENSON

A new space race has begun, and most Americans are not even aware of it.

This race is not merely between two nations jockeying for political prestige or military power. This new race involves the whole human species in a contest against time. All of the people of Earth are in a desperate race against global disaster.

The end of civilization is in sight, now, in the smoking streets of Tehran and Belfast and Miami, in the starving masses of the Sahel and Cambodia, in the nuclear arsenals and imperial ambitions of many nations. Only by raising our sights above the immediate problems of the moment, only by reaching outward into space itself, can we avert the coming worldwide collapse of civilization and the deaths of billions.

Problems press in on us from all sides, and we are so busy dealing with what is urgent that we have lost track of what is important. We have become short-sighted, thinking only of the immediate crisis. We perceive the future as something quite remote from us; then we are shocked by the sudden changes that the future brings.

If we expect to see the next century, or to have our children survive in it, then we must adjust our vision for the long view. We must begin building the twenty-first century now. Today. We have already wasted too much precious time. Very little time is left in which we can act.

To save the Earth we must look beyond it, to interplanetary space. To prevent the collapse of civilization and the end of the world as we know it, we must understand that our planet does not exist in isolation. Our Earth is part of a Solar System that is incredibly rich in energy, in natural resources, in all the wealth and raw materials that we need to build a flourishing, fair, and free global society.

This new space race, in reality, is a crucial struggle against humankind's ancient and remorseless enemies: hunger, poverty, ignorance, and death.

We must win this race, for one brutally simple reason: survival.

2

Future One

The year is A.D. 2000.

There are forty million unemployed in the United States of America. Ever since the economic Collapse of the late 1980s, the nation's standard of living has plummeted until now, at the turn of the century, most Americans live no better than their forefathers did two centuries ago.

Canada has closed the long-unguarded frontier between the two nations to prevent "drybacks" from sneaking into the country to seek work. They have caused a huge upsurge in crime and welfare costs, the Canadian government complains daily. Labor gangs are recruited in most major cities, where criminal offenders are given the choice of jail or work gangs. In the Southwest, such labor battalions are often shipped to Mexico. None has returned.

Some blame the Collapse on the energy crisis. For more than a decade the US paid out its hard-earned treasure for foreign oil, while neither conserving its own resources nor developing new sources of energy. Now, unable to buy foreign oil, without the money even to mine the vast reserves of coal under its own soil, the US sinks deeper into poverty and despair.

The riots and insurrections of the 1990s have quieted, at least.

In the ghostly cities and dusty towns there are hardly any youths to be seen. Conscripted into the Army, the Conservation Corps, the labor gangs, they are no longer free to rage and burn. Ironically, one of the largest labor contingents in New York has been set to work repairing the partially destroyed Statue of Liberty.

A Presidential election is coming up this November, and the word has gone out that every eligible voter will vote. Not that it makes much difference: Both candidates have been preselected by the apparatus in Washington and okayed by Moscow, although which one has actually been selected to win the race is being kept a secret, for entertainment reasons.

3

Millennial Angst

> The earth is degenerating these days. Bribery and corruption abound. Children no longer mind parents, every man wants to write a book, and it is clear that the end of the world is at hand.
> —ASSYRIAN TABLET,
> CIRCA 2500 B.C.

Ask a writer where he got the inspiration for his book and usually he will respond with a blank stare. But I can tell you exactly when and where this book began.

Faneuil Hall, Boston. July 20, 1979.

I had been invited to give a speech in Faneuil Hall on the tenth anniversary of the Apollo moon landing. Naturally enough, the speech was to be about the future of the American space program. I was excited about it. Faneuil Hall is where the angry Massachusetts colonists donned their Indian disguises just before they went out to the docks and dumped British tea into Boston Harbor. Birthplace of the American Revolution, starting point of the Boston Tea Party, and I was to speak there about the new revolution that is going to take place in space.

It was a Friday afternoon. Earlier in my life I had lived in

Boston, but I had completely forgotten how impossible Boston traffic can be, particularly downtown on a summer Friday afternoon. We got to within a block of the hall and then bogged down in an immense honking, growling traffic jam. I thanked my driver for courage beyond the call of duty and bolted from the car, my tray of 35 mm slides in one hand, and picked my way through the stalled automobiles.

Breathless, I arrived inside the hall about a minute and a half late. It was like stepping back into the eighteenth century. Faneuil Hall had been restored to its original austere, white, Colonial decor.

And there were fourteen people in the audience.

The night before I had spoken to a crowd of several hundred at the McDonall Planetarium in St. Louis. But here in the birthplace of the American Revolution only two more than a dozen had shown up to help celebrate the tenth anniversary of the lunar landing.

I went through my talk automatically, showing the slides and commenting on them by rote. A family of five wandered in off the street, took a row of seats at the rear of the hall, and after five minutes drifted out again.

In the back of my mind I began to recall where I had been ten years earlier, on the night that Neil Armstrong and Edwin Aldrin first set foot on the Moon. I was in Mike Pratt's family room, surrounded by his wife and children and my own, out in the suburbs of Boston's South Shore. We all watched every moment of that historic event, spellbound. When Armstrong actually stepped out onto the lunar surface and made his slightly pompous statement about "a giant leap for mankind," we popped a bottle of champagne. I still have the cork.

On that night, July 20, 1969, I naïvely assumed that since we had successfully reached the Moon, the next ten years would be packed with solid achievements in space. It seemed clear to me that we had the tools and the team to begin to use the space environment for the betterment of life on Earth. Weather and communications satellites were already realities; I looked for ward to permanent orbiting stations where people could create new industries, make new scientific discoveries. I foresaw cities growing in space, where men and women of vision and courage could explore this New World and build new lives for themselves.

As I looked out on that skimpy audience in Faneuil Hall, July 20, 1979, it came home to me with crushing force that we had wasted ten years.

Ten wasted years.

Instead of a permanent space station we lofted Skylab — a jury-rigged vehicle cobbled together from the leftover pieces of the murdered Apollo program. Instead of offering a permanent way-station in orbit, Skylab plunged to a fiery death because NASA had neither the funds nor the equipment to save it from plummeting back to Earth. Instead of using space to create new industries, new discoveries, new jobs and profits, we have allowed the space program to bog down. Our highly trained team has dispersed, our hardware is years out of date. Our one major program — the reusable Space Shuttle — was dangerously underfunded and compromised in design, and as a result lagged years behind its original schedule.

We have achieved some stunning successes in space over the 1969–79 decade. Thanks to the technology and skills developed on the Apollo program, we were able to send unmanned exploratory spacecraft such as the Pioneers, Voyagers, and Vikings to the other planets of the Solar System. During our Bicentennial Year we landed two Viking spacecraft on the surface of Mars. (By a coincidence, Viking I touched down on the Martian surface on July 20, exactly seven years after the first Apollo landing.)

But although these unmanned explorers have told us much about the other planets of the Solar System, they have given us nothing of immediate practical value. The American public is keenly aware of this. Magnificent scientific achievements that they are, they don't put any bread on the table. Viking's touchdown on Mars had nowhere near the impact to the average

American as a tax cut, or the loss of a job, or the worldwide crises that blare out of the headlines every day.

Ten wasted years.

By 1969 we had forged a tool that allowed us to reach any part of the Solar System that we desired. That tool was a complex mix of people and machines, knowledge and skills, technology and spirit.

During the decade of the seventies we should have used that tool to make life better on Earth, to help solve the crises of energy, ecology, and economics that have plagued us.

But even as President Richard Nixon was congratulating the returning Apollo astronauts for their magnificent achievement, his aides in Washington were slashing the space budget and making it certain that we would not be able to use our space technology in any practical, useful manner. And we let them get away with it. You and I, the taxpayers. We have allowed the machines to rust and the skilled team of people to scatter, bitter, confused, heartbroken.

On the lawn just inside the main gate of the Johnson Space Center, near Houston, lies a complete Saturn V rocket booster, stranded upon the sun-scorched grass like a beached whale, like the remains of a dead dinosaur. That Saturn V was built to carry astronauts to the Moon. It is not a model or a mockup, it is the real rocket. It cost us $250 million to build. Useless now, bleaching in the sun. That is the most fitting symbol I can think of for those ten wasted years.

And during that decade, did we use the money not spent on space to solve our problems? Our nation has sunk into crippling inflation, unemployment, loss of military strength, loss of national prestige, and a discernible lowering of our standard of living. Productivity increases among American workers were negative in 1979, and again in 1980. Where once the US had the highest per capita gross national product in the world, now we rank seventh. And sinking.

All of these problems could have been alleviated, perhaps averted altogether, if we had pursued a vigorous space program. Is it a coincidence that our economic slide began when we started whittling away at the space budget?

Now it is the decade of the eighties. Our problems grow worse, our ability to face them and solve them grows weaker, and the vultures are beginning to circle over our heads.

All of this went through my mind as I recited my speech to those fourteen patient and somewhat puzzled people in Faneuil Hall.

That is where and when this book began.

*

It is fashionable to predict the end of the world as a new millennium approaches.

A thousand years ago, as Christian Europe faced A.D. 1000, millennial fever swept across the land. People foresaw the Second Coming of Christ, the day of judgment, the end of the world. Repentance of sins and harsh penance — including voluntary crucifixions — were not uncommon.

Today, as we face A.D. 2000, the doomsday predictions come from economists and computers, from political sages and military analysts, from earnest young people who despair of protecting our natural environment and from disillusioned old people who see the collapse of everything they spent their lifetimes building. We have our choice of catastrophes, as a glance at any newspaper or television documentary will tell you: the proliferation of nuclear weapons, the growing instability of the Middle East, confrontation between the United States and Soviet Russia, ecological disasters, volcanic eruptions, earthquakes, the melting of the ice caps (which would drown most of civilization), a new Ice Age — disasters real and potential surround us.

And as problems and crises multiply there is a growing temptation to strike back, to find the simple, easy solution, the one quick stroke that will make everything well again. Find an enemy: In Iran, it was the Shah; to antinuclear protestors, it is the power industry; to a Northern Irish Catholic, it's the Protestants; to an Ulsterman, it's the Catholics; to the terrorists, the mystics, the rebels, the kids in the streets, it's Them. The enemy. Get rid of Them and the world will be all right again.

Perhaps this is the greatest danger of them all, this descent into irrationality that can destroy not only our democracy, but our very civilization.

Western civilization has been built on the foundation of reason. It is no coincidence that the Founding Fathers of America were rationalists, well acquainted with contemporary science. Jefferson and Franklin were even something of scientists themselves.

For better or for worse, science has transformed the world. For the most part, our lives have been enriched and made better by science and its offspring technologies. Certainly these benefits have not come without costs, but I for one would prefer to live as I do today rather than as my forebears did in Calabria only a few generations ago.

We are beset by the armies of ignorance, the irrational mobs who run through the streets, shutting their minds to reason, shouting, "Death to the Shah" or "No more nukes" simply because some authority figure told them to do so. They might as well be shouting, "Let Him be crucified," or even "Seig heil!"

The only real enemy of the people is ignorance.

And the only real problem we have is ourselves: There are too many of us. We have outgrown this planet.

4

The Flat-Earth Fallacy

The real problem . . . is not too many poor
people; it's too few rich. Get the distinction?
—KELVIN THROOP

All the disasters we face, from nuclear war to ecological collapse
to the tide of irrationality, have one factor in common: population pressure. There are more than four and a half billion people living on Earth today. (And, as I write these words, two men
living in orbit aboard the Soviet space station Salyut 6.) By the
turn of the century, there will be at least six billion people.
Perhaps seven.

The political, social, ecological, and emotional strains we feel
today all stem from the fact that four and a half billion mouths
must be fed every day. And every day, more than 200,000 new
babies are added to the problem. Each of these human beings
requires food, living space, energy, clothing, housing. To provide these necessities, we are consuming our planet's natural
resources of fuels, metals, minerals, timber, farmland.

Lester R. Brown, president of the Worldwatch Institute, has
warned that we are rapidly destroying the biological resources
on which we all depend. The human race is almost totally depen-

dent on four worldwide biological "systems," in Brown's view: fisheries, forests, grasslands, and croplands. All our food and most of our raw materials (except for minerals and fossil fuels) come from these four biological sources. Biological systems are renewable; they automatically renew themselves, in fact, because they are living, dynamic systems. But there are limits to how hard we can press such systems without destroying them. Put too much pressure on a biological system, and, instead of renewing itself, it will self-destruct. We are harvesting the world's fisheries, forests, grasslands, and croplands faster than they can replenish themselves, leading to a situation where, in Brown's words, "fisheries collapse, forests disappear, grasslands are converted into barren wastelands, and croplands deteriorate."

The southward creep of the Sahara Desert and the resulting famines in the Sahel region are one manifestation of human overpressure on the Sahel grasslands.

The decreasing returns from croplands, despite constantly increased doses of fertilizers and insecticides, are another result of human overpressure. So is the stripping of the tropical rain forests of South America. So is the collapse of the fisheries off Peru, Cape Cod, the Norwegian coast. Where Odysseus once plucked fish out of the wine-dark Aegean, Greek fishermen now find a barren sea.

Brown is an economist, and he states drily, "The condition of the [world] economy and of these biological systems cannot be separated." Human population pressure is driving us to the destruction of the biological systems on which we depend. Yet what else can we do? There are more than 200,000 hungry new mouths to feed every day.

Technical experts have argued for years now about how many people planet Earth can actually support. Six billion? Ten? There are those who say our world has the physical resources to accommodate a human population of twenty billion or more. But others say that anything above a few hundred million is overpopulating this planet. Whatever the experts believe, the painful truth is this: With four and a half billion people

on Earth, the social fabric of human civilization is tearing apart.
The so-called energy shortage is merely the first tip of a vast
iceberg. Our global supplies of petroleum are growing short.
That's not a terribly frightening matter, in itself. Petroleum is
only one source of energy out of many that are available to us
— ranging from coal and uranium to sunshine and wind power.
Yet, because of the growing scarcity of petroleum, wars have
been fought, national economies all around the world are tottering, millions of workers are unemployed, the global mone-
tary system is in danger, small nations find it harder than ever
to feed their people, and the superpowers are returning to the
tactics of nuclear saber-rattling.

What will the world be like in the next decade, when there
are six or seven billion of us, when petroleum is even scarcer
and more expensive than it is now, when there are critical
scarcities of such truly vital resources as copper, phosphates,
and arable land on which to grow food crops?

In 1972 an international group of scientists reviewed these
problems and came to the conclusion that the world is heading
for global disaster, a collapse of civilization on a scale not seen
since the Roman Empire self-destructed nearly two thousand
years ago.

Billions of human beings will die. The survivors will be reduced to a medieval standard of living, or worse.

Superdisaster.

The scientists' warning of doom was publicized in a book
titled *The Limits to Growth* (Universe, 1974), written by the
team from the Massachusetts Institute of Technology who actually derived the forecasts from computer analyses of the world's
economic and population trends. The work was done under the
sponsorship of the Club of Rome, an international association of
industrialists, scientists, educators, and civil servants. *The Limits to Growth* was hailed and vilified. Environmentalists and
others who had been warning of the follies of industrialized
civilization praised the report as scientific proof that unless we
change our ways, civilization is doomed. Other scientists, indus-

trialists, and politicians damned the book as alarmist, inaccurate, and narrow-minded.

In 1977, President Jimmy Carter assigned the Council on Environmental Quality and the Department of State to make a report to him on "the probable changes in the world's population, natural resources, and environment through the end of the century." The report, released in July 1980 and titled *Global 2000 Report,* reconfirms the findings of the Club of Rome's *Limits to Growth.*

"If present trends continue," *Global 2000* warns, "the world in 2000 will be more crowded, more polluted, less stable ecologically, and more vulnerable to disruption than the world we live in now." In particular, *Global 2000* points out that between 1975 and 2000, the world's remaining per capita petroleum resources will dwindle by 50 percent. Per capita water supplies will shrink by 35 percent merely because of population growth alone. Forty percent of the forests in Third World nations will have been cut down, and desertification will claim "a significant fraction of the world's rangeland and cropland" by 2000.

Prices will be higher. "In order to meet the projected demand, a 100 percent increase in the real price of food will be required."

"The world will be more vulnerable both to natural disaster and to disruptions from human causes . . . The tensions that could lead to war will have multiplied."

There are many shortcomings in both *The Limits to Growth* and the *Global 2000 Report.* Yet the simple fact remains: As more and more people use up more and more of the Earth's natural resources — something's got to give.

All this begins to sound like Malthus in a computer. Thomas Malthus, the English economist and historian, first propounded in 1798 that population growth always tends to outstrip all possible growth of food production, and therefore starvation and poverty are the natural order of human existence.

The Malthusian point of view not only served as a theoretical justification for the cutthroat exploitation of the poor that was a major feature of nineteenth-century capitalism (and as a result gave birth to Marxism), but Malthus's gloomy outlook also helped to give economics the reputation for being "the dismal science."

Malthusian or not, *The Limits to Growth* and *Global 2000* paint a bleak picture. While many arguments have been mounted against them, no one can refute their basic point: If human population continues unabated we will use so much of the Earth's natural resources that civilization will collapse and billions of people will die.

The conclusion reached by these studies is strikingly clear: We cannot continue to have a growing society.

Given our free choice, we will always opt for more industrial production, no birth control, faster and faster consumption of natural resources, all of which lead to disaster. We must stop our drive for constant growth and stabilize our population, production, and consumption.

How?

In the industrialized nations there is already considerable sentiment toward conservation of natural resources, control of growth, limits to population. Birthrates have fallen in the United States and Western Europe, for example, close to the Zero Population Growth level.

But these are the rich nations, the "haves." Rich people can afford to worry about the quality of life and of the environment. Poor people can't.

The "have nots" are either unable or unwilling to limit their population growth, their industrial production, the growth of their national economies. "Genocide!" they cry, whenever a rich, Western, white nation suggests that a poor, undeveloped, nonwhite nation slow its growth rate.

All around the world, from Detroit to Djibouti, the struggle rages between those who say, "Enough!" and those who say, "I'm hungry."

Historically, there have been two classic ways to control population growth.

One: The Four Horsemen. Let famine, pestilence, war, and death do their number. The other: Raise the standard of living. Rich people have fewer children than poor people. If you don't believe that, take a trip to Mexico or Egypt or Bangladesh or Zaïre. Or the South Bronx or the black or Hispanic ghettos of any city in the United States.

The average white family in the United States has between two and three children. The average family in Mexico, nine. That's *average*. I know one nationally prominent television personality of Puerto Rican ancestry who had twenty-one brothers and sisters. "You should see our family reunions," he says, grinning.

It is easy for the rich to tell the poor that if they would merely control themselves for a while, the world's problems will be solved. But try saying that in a village in Pakistan or Peru. Try saying it in Harlem, or even in my old neighborhood in South Philadelphia's Italian section. The poor peoples of the world will not, cannot, voluntarily control their population growth. Their cultures, their religions, their deepest instincts work against the very idea of birth control, abortion, sterilization.

Can we control worldwide population growth by nonvoluntary means? By passing population laws and enforcing them? Indira Gandhi tried it in India and was thrown out of office. The government of China is attempting to do it more subtly, linking job opportunities and socioeconomic incentives to low birth rates. Most other national governments have not and will not make the attempt.

Many science fiction scenarios have been written in which a world government enforces strict population laws. Almost all of these stories are dismal. They inevitably end either in the death of the hero and/or heroine (as in Aldous Huxley's *Brave New World*) or in the totally unconvincing overthrow of the dictatorial regime by the selfsame hero and/or heroine.

If you can envision a world government powerful enough to make population laws stick, you have imagined a truly crushing dictatorship. Big Brother will be watching *you*, in your bedroom, your bathroom, and everywhere else. It might be possible, using modern technologies of electronics, drugs, mass media, computers, and coercion to enforce such a dictatorship. But we will have achieved population control at the cost of freedom. We will exist, but at the whim of an overpowering authoritarian rule.

Better Red than overbred? There must be a happier way.

The only viable alternative seems to be to increase the standard of living for everybody, especially the poor. Make them rich enough so that they don't need to have nine children per family.

But despite the enormous efforts by the developing nations to increase their gross national products, the very population growth they are trying to stem always wipes out whatever economic growth they manage to achieve. What good is it to increase the GNP by two percent when the number of mouths you must feed is increasing by four percent? This is why the poor nations of the world have turned their hungry eyes on the rich.

"Share what you have with us," they cry, with growing anger in their desperate voices.

"But if we do," say the rich, "then we will all become poor." And there is truth in that.

So we all sit in the same pot, simmering away, while the tensions and conflicts among the nations grow worse with each passing year.

And every day more than 200,000 new babies are born.

And every day the world's supply of natural resources dwindles.

The poor need what the rich have. The rich know that there simply is not enough to go around.

All of this is taking place in the midst of the most gigantic supply of natural resources that any human being can imagine. All the resources we need to make every human being on Earth as rich as an emperor lie within the grasp of our fingers. Yet most people have no inkling of this fact.

We are starving to death in the midst of plenty.

The Malthusian dilemma, the dismal conclusions of *The Limits to Growth* and *Global 2000 Report,* all share a common fallacy. These analyses of the human condition tacitly assume that the Earth is flat, that there are no other worlds in the universe, that we are stuck here on the ground forever.

They are wrong.

The Flat-Earth Fallacy is as pervasive as it is perverse. We all share it, to some extent. Even though we have lived through the so-called Space Race and have seen men on the Moon and robot explorers on Mars, we still tend to think that we have only the surface of this one planet Earth on which to solve our problems.

A typical example of this fallacious thinking is in Richard J. Barnet's book, *The Lean Years* (Simon and Schuster, 1980). Barnet concludes, "Both capitalism and socialism must confront the 'tragedy of the commons' noted first by the nineteenth-century Malthusian William Forster Lloyd . . ."

What is the "tragedy of the commons"? According to Barnet:

> Men seeking gain naturally desire to increase the size of their herds. Since the commons [grazing ground] is finite, the day must come when the total number of cattle reaches the carrying capacity; the addition of more cattle will . . . eventually destroy the resource on which the herdsmen depend . . .

Since the commons is finite. That's the fallacy. It isn't. Not anymore.

Our planet is part of a solar system that is very hostile to human habitation, yet incredibly rich in energy and natural resources. It may be centuries before large numbers of people can live in space. But we can, today, send a small number of men and women into space to begin to tap the resources waiting there for us.

We do not want to export people. We do want to import wealth.

Lift up your faces. The Sun smiles upon you. The Moon beckons.

5

Getting There

> Every revolutionary idea . . . seems to evoke
> three stages of reaction. They may be
> summed up by the phrases: (1) It's com-
> pletely impossible . . . (2) It's possible, but it's
> not worth doing; (3) I said it was a good idea
> all along.
>
> —ARTHUR C. CLARKE

When Armstrong and Aldrin set down on the Moon's dusty
surface, space enthusiasts all around the world thought that
their mission had been accomplished. We had fought our way
past Clarke's first two phases, and everyone was congratulating
each other on a job well done.

Even Vice President Spiro Agnew, not known for his advanced views on research and technology, was quoted as proclaiming, "On to Mars!"

But within a year the *Boston Globe* was running an editorial
cartoon showing the chief of NASA as a street-corner panhan-
dler, begging for funds in the snow. I clipped the cartoon from
the paper and hung it on my office wall with the caption, "Remember, Columbus was brought back to Spain in chains."

Today, more than a decade after Apollo's stirring success, we

seem to be back on Square One. Tell someone about Solar
Power Satellites solving our energy crisis, or lunar mining operations to feed factories in orbit, or energy beam weapons in
satellites shooting down ballistic missiles, and you get Clarke's
phase-one reaction: "It's completely impossible!"

We have the knowledge, the skills, the technology, and
trained people to begin to utilize space to solve our problems
here on Earth. But do we have the vision to pick up this magnificent tool and use it wisely? Do we have the heart even to try?

When I was seven years old, growing up among the row
houses of South Philadelphia (you saw some of them in the film
Rocky), one of our neighbors was a marvelous old salt who had
served in the Navy in World War I. Many a muggy summer
evening, when the whole neighborhood would sit out on the
front steps trying to catch a vagrant breeze, he would call me
as I walked past his house. Naturally I would turn back and
come to the base of his steps; all kids obeyed all adults instan-
taneously in my old neighborhood. Or else.

"How far would you be if I hadn't called you?" he would ask
me.

It was a tough question for a seven-year-old to figure out.

How far would we be today if NASA's fiscal guts hadn't been
ripped out by a succession of know-nothing Administrations
and shortsighted Congresses? Would there be human footprints
on the red sands of Mars? Would we be reopening factories and
steel mills and navy yards to step up production for the growing
demand for space hardware?

Would we have an unemployment rate of almost ten percent? Or double-digit inflation?

Would we have had to endure fifty-three hostages sitting in
Iran for month after month, or Soviet troops in Afghanistan?

Tough questions. I don't pretend to know the true answers to
them. But it seems to me that our strength as a nation is based
on the strength of our economy. And space operations can be
a mighty source of strength to our economy.

More than that. Our freedom as individuals is based not

merely on the nation's political power, but on the real wealth that we possess as individuals. A peasant who toils from dawn to dusk, only to collapse exhausted into sleep to await the next dawn, is not as free as a farmer who uses modern mechanized farm equipment. Technology has always led, throughout history, to increased individual freedom. In our time, space technology is the cutting edge of technological advances. Increases in human freedom are inextricably bound to advances in technology, especially space technology.

Why aren't we using this magnificent technology to solve our problems and advance the cause of human freedom? How did we fall down so badly?

Compared to where we could be in space, it's as if the Apollo project never happened. We are no closer to the Moon today than we were in 1960. We have no more boosters capable of taking astronauts into deep space; those that we haven't used have been scrapped, or turned into museum displays. All the knowledge and skills that we built up during the Apollo years are in hibernation.

How did we permit this to happen?

It goes back to the Space Race. Make that Space Races. Plural. There were two of them.

As all things do, it began with the poets.

In fact, you could say it started with Cyrano de Bergerac. The hero of Edmond Rostand's famous play was a real person who lived in seventeenth-century Paris. He actually was a famous swordsman and poet (and, unlike the character in the play, quite a womanizer). He wrote satirical stories that today we would call science fiction. Cyrano was the first writer, actually, to hit upon the idea of using rockets to travel beyond the Earth.

It may seem self-serving to describe science fiction writers as poets, but in the largest sense of the word, the romantic traditions of poetry have been carried forward in our times by the writers who have pictured a brighter, saner, grander world than the one we live in — and these are the writers of modern science fiction.

There is a lot of bad science fiction, to be sure, and even some of the best of it can be gloomily dystopian. But during the 1930s and '40s, when the rest of the world was tooling up for war, science fiction writers were showing utopian visions of future worlds in which the human race had reached out into the Solar System and even farther, toward the stars.

World War II ended with a double spasm of destruction. In Europe, the V-2 rocket: unstoppable by any conceivable defense. In Asia, the nuclear bomb: devastation so awesome that it made war unthinkable even for the *Bushido*-trained Japanese militarists.

The V-2 was the product of Wernher von Braun, who was introduced to notions of rocketry through science fiction, and brought as a teen-ager to the German Society for Space Travel by the writer Willy Ley.

Many of the engineers and scientists who worked on the Manhattan Project read science fiction and even saw a working description of their Fat Boy bomb appear in the pages of *Astounding Science Fiction* magazine months before the Trinity test gave the world its first mushroom cloud, at Alamogordo. The FBI investigated the "security leak," and found that writer Cleve Cartmill had based his fiction on prewar knowledge and accurate guessing.

When long-range rockets and nuclear bombs appeared in the real world, science fiction gained a new respectability. People who had always considered science fiction to be juvenile pulp literature were willing to listen.

And think.

Arthur C. Clarke combined the imagination of science fiction with a solid background in physics and engineering to invent the idea of communications satellites. To this day, Clarke has never received one penny from his idea, which prompted him to write a charming essay titled, "How I Lost a Billion Dollars in My Spare Time."

Robert A. Heinlein, already the dean of American science fiction by the time the war ended, turned one of his novels into the screenplay for George Pal's 1950 movie, *Destination Moon*. Clarke's solid yet imaginative and often moving prose convinced a wide range of readers that space flight was possible. *Destination Moon* played to standing-room-only audiences all around the world. The comparison may be odious, but it was the *Star Wars* of its day.

The American public began to realize that we could reach the Moon if we wanted to.

But why should we want to?

It is difficult to realize today, but merely a generation ago the phrase "fly to the Moon" was the ultimate expression of impossibility. For example, in 1948, political pundits compared Harry Truman's chances of winning re-election with his chances of flying to the Moon.

Apollo was less than a dozen years away and Truman won the election. So much for the foresight of political pundits.

Arthur Clarke did not restrict his work to writing. He served as chairman of the British Interplanetary Society, one of the pioneer groups of space enthusiasts whose members formed the nucleus of the newly emerging British aerospace industry.

In 1949 Clarke and a few of his BIS fellows convinced an American physicist, S. Fred Singer, that satellites would be useful for scientific studies of the upper atmosphere. Singer, who was serving as a science attaché with the US embassy in London, had already been involved in rocket work, where captured German V-2's and American rockets had been used to study cosmic rays at altitudes of up to 100 miles. Over pints of beer, the BIS crew and Singer drew up a program for a small artificial satellite called MOUSE — Minimum Orbital Unmanned Satellite of Earth. MOUSE never came to fruition, but it led to other things.

While those things were developing, Clarke was asked to produce a symposium about space flight by the Hayden Planetarium in New York. He wrote to Dr. Harry Wexler, chief of research for the US Weather Bureau, and asked Wexler to give a paper on the use of satellites for meteorological work. Wexler replied that satellites would be useless for weather research. At this point, Clarke showed the difference between talent and genius. He invited Wexler to present a paper on why satellites would *not* be useful to meteorologists. Wexler agreed to that. Being an honest man and a good scientist, Wexler found, once he looked into the matter, that he had been wrong. Satellites would be enormously helpful to weather researchers and forecasters. He became a space enthusiast. The facts convinced him.

Meanwhile, back in the Pentagon, unsettling news was percolating up through the chain of command. The Russians were flight-testing big rockets. Missiles. Big enough to carry nuclear warheads over thousands of miles and hit cities in America. The term ICBM — Intercontinental Ballistic Missile — entered the vocabulary of the military. And the White House.

Von Braun, of course, had come to the US after the war, together with more than a hundred of his engineers and technicians from the Peenemünde rocket development center. Why were the Russians flying big rockets while the US Army kept von Braun down in Redstone Arsenal, working on upgraded versions of the V-2?

Because of a bad interpretation of scientific evidence.

Immediately after the war ended, the US government asked Dr. Vannevar Bush to assess the chances of building nuclear-armed intercontinental missiles. Bush had been director of the Office of Scientific Research and Development during the war, overseeing the entire American wartime R&D operation.

Looking over the scientific evidence, Bush concluded that nuclear bombs were so large that no conceivable rocket could carry them. So the US moved to develop very large bomber aircraft, such as the B-36 and B-52, and ignored rocket development. The Russians, looking at exactly the same evidence, came to the conclusion that a nuclear-armed missile would have to be huge, compared to a V-2. So they began to design and build rockets big enough to do the job.

Bush had no faith in rocketry and made the crucial error of

believing that the conditions of the moment would never change. Rocketry development in America was stunted as a result, while rocketry development in the Soviet Union was accelerated — because of the same evidence.

Within a few years, of course, conditions changed. Both sides figured out how to make nuclear weapons smaller without sacrificing explosive yield: more bang for the pound, or kilogram.

In the late 1940s, Bush confidently wrote, "There has been a great deal said about a three-thousand-mile rocket. In my opinion such a thing is impossible for many years. I think we can leave that out of our thinking." It was a self-fulfilling prophecy. When an official in the lofty position Bush held makes a negative prediction, the bureaucracy under him automatically rejects any ideas or suggestions that run counter to it.

Sputnik was less than ten years away. As Arthur Clarke has pointed out, "When a distinguished but elderly [in this context, meaning "over thirty"] scientist says that something is possible, he is almost certainly right. When he says it is impossible, he is very probably wrong."

By the early 1950s it became quite clear that the Russians were flight-testing long-range missiles. In deepest secrecy, the United States began an all-out crash program to catch up to the Russian lead. This was the first Space Race, the one that was hinted at later, during the 1960 Presidential election campaign, when the term *missile gap* became a catch phrase.

In 1956, while we were struggling to catch up with the Russians, the Hungarians threw off their Communist government and pleaded with the West for help. Despite the election-year rhetoric of President Dwight Eisenhower and his Secretary of State, John Foster Dulles, to "free the captive peoples of Eastern Europe," neither the US nor any nation of Europe came to the aid of the Hungarians. We all stood idly by while Soviet tanks crushed the Freedom Fighters of Budapest.

Little attention was paid by Western media to the fact that Nikita Khrushchev, the Soviet Premier, had sent word to every capital in Western Europe that if they tried to intervene in

Hungary, Soviet rockets would level their cities. The Russians had operational missiles by then, capable of reaching any city in Europe.

All the while, von Braun was an employee of the US Army, constrained to work on developing an upgraded version of the V-2 named the Redstone, in honor of the arsenal in Alabama where he was based.

Science fiction writers were trying to further the idea of peaceful space flight, and von Braun teamed with painter Chesley Bonestell to produce a series of stunning illustrated articles about space flight in *Collier's* magazine. Many imaginations were stirred by von Braun's matter-of-fact prose and Bonestell's magnificently detailed paintings. But the magazine folded.

In the midst of all this, S. Fred Singer returned from England afire to use satellites for peaceful scientific research. He quickly found that there was a huge hush-hush rocket program racing full throttle in competition with the Russians. (And, in the manner of the Pentagon in those days, the three armed services were also in a cutthroat competition with one another.)

The scientists advising Eisenhower agreed with Singer's ideas, but for a variety of reasons the White House wanted to keep military rocket developments strictly separated from civilian rocketry projects. Singer and his friends in the BIS had picked as inoffensive a name as they could for this scientific satellite proposal: MOUSE. When it became clear that the government would not appropriate funds for the project, Singer proposed a more modest effort, which was soon dubbed MINI-MOUSE.

Enter now the international scientific community. Geophysical scientists all over the world had organized a global cooperative research effort called the International Geophysical Year, or IGY. The IGY's "year" was actually eighteen months long, starting July 1, 1957, and lasting to December 31, 1958. Geophysical scientists of all disciplines and nations, including the USSR and Eastern Europe, would cooperatively study every aspect of the Earth, the Sun, and the interplanetary medium that linked the two. Thanks to the effort of Singer and others, the American participation in the IGY was expanded to include a program of orbiting several scientific satellites. The Russians also announced that they would put satellites into orbit, but hardly anyone in America paid attention to Soviet propaganda. Most Americans thought of the Russians as tenacious peasants, not space scientists.

The American satellite program was called Vanguard. It was a "peaceful" operation, which meant that the Eisenhower Administration forbade Vanguard's use of military rockets. Von Braun's Redstone was ready to be used, but that was an Army rocket. Vanguard was to be developed from peaceful boosters, such as the already-existing Viking and Aerobee rockets, which were used to probe the upper atmosphere. The political decision-makers in Washington had not yet tumbled to the fact that a booster is a booster, just as a truck is a truck. What determines the degree of its belligerency is what it carries as its payload.

To confuse the picture even further, the government had to turn the management of the Vanguard Project over to one of the military services. There was no civilian space agency, and the only government organizations with any experience in handling rocketry programs were in the military. After a typical interservice struggle, the Navy won the assignment. The Office of Naval Research ran Vanguard, and the Naval Research Laboratory built the gold-plated (literally) Vanguard satellites. It was a shoestring operation, despite the satellite's golden coat. (Among its many useful properties, gold is an excellent thermal conductor.) Most of the nation's rocketry talent and funding was in the multi-pronged ICBM crash program. Vanguard got leftovers.

And beginners. One of the Junior Technical Editors on Vanguard was a 24-year-old refugee from newspapering who had no formal scientific training and no technical knowledge of rocketry: me. I talked myself into the job with the Glenn L. Martin Company (now Martin-Marietta Corporation) on the strength of some writing ability and a lot of enthusiasm. Martin built the Vanguard launching rockets in its vast factory just outside Baltimore. That factory, by the way, no longer builds aerospace hardware.

The stars in my eyes when I joined Vanguard were quickly washed away by sweat, and then by tears of frustration. The Vanguard engineering team was ensconced in the loft atop one of Martin's manufacturing buildings. Summers are hot in the Baltimore area. And muggy. Our desks were located above machinery that used molten aluminum. On a summer morning, desk-top temperatures would be over 90° F before 8 A.M. Pigeons nested in the girders above us, and the desks were frequently spattered by their strafing runs. Oh, it was *swell.*

Esprit de corps was very high, though, and we all laughed at the environmental difficulties. Vanguard was well into its testing program when Sputnik I roared into orbit on October 4, 1957. The shock wave that raced around the world was incredible. Western media, especially, were hysterical. The Russians had at last made it painfully clear even to the most closed-minded pundits that they did indeed have rockets capable of depositing hydrogen-bomb warheads on any city on Earth. The Space Race that the public knew about had begun.

Actually, in the race that really counted, we were catching up pretty quickly. Atlas and Titan ICBMs were coming along, as was Admiral Hyman Rickover's Polaris submarine missile system. Minuteman ICBMs were also being planned. But that was mostly hidden from public view. By the time John F. Kennedy came into the White House, in January 1961, he found that the "missile gap" had been reversed: We had more ICBMs and submarine missiles in place than the Russians did.

But meanwhile a highly visible space race was going, and we were behind — badly behind. Vanguard was to launch a 20-pound satellite. Sputnik I weighed 200 pounds. In November 1957 the Russians orbited Sputnik II, which weighed half a ton and carried a dog, Laika, aboard.

The US Navy did not panic. Not then. But when von Braun announced that he was prepared to launch a jury-rigged satellite on a modified version of his Redstone rocket, the Navy went berserk. Von Braun worked for the *Army!*

Vanguard Test Vehicle No. 3 was transformed into a satellite launching vehicle, on orders from the Navy. No matter that this was the first Vanguard rocket to have all three stages fitted with live engines. No matter that the General Electric Company's first-stage engine was marginal in its performance, at best, and had already been shipped back to GE's plant at Schenectady several times for rework. The Navy brought a special two-pound test satellite, quickly dubbed "the grapefruit," and announced to the world that we would fling it into orbit.

With live television coverage.

The rocket got four feet off its launch stand and exploded. America was humiliated. The following March, on St. Patrick's Day, a Vanguard satellite did achieve a successful orbit (with a St. Christopher's medal welded to its guidance section). But by then it was too late. Vanguard, and American rocketry in general, had flopped before a worldwide audience.

The rest is history.

Von Braun put the first American satellite, Explorer I, into orbit in January 1958. But that only seemed to accentuate America's lack of ability: It took the Germans to get us into space.

The Russians ran off a string of spectacular "firsts," including the first man in space — the late Yuri Gagarin.

Then John Kennedy came into the White House and, for a whole hatful of political, economic, and prestige reasons, convinced the Congress and the American people that we should aim for the Moon.

It was sound strategy. It leapfrogged the Russians. In the old infantry-tactics sense, we "took the high ground." By going to the Moon, we forced ourselves to develop a space technology of hardware and people that could take us anywhere in the Solar System.

The science fiction writers were delighted, at first. They had sung their songs and set the stage so that the people were ready to support flights to the Moon. But then the engineers came in, said a curt "thankyou," and took over the operation for themselves. No science fiction writers needed anymore. It was a time for engineering, not poetry. Numbly, the writers stepped aside and watched.

Engineers are lousy at public relations. NASA created a stilted, stultified mythology about the astronauts that made them appear to be a cross between Tom Swift and George Washington. As Tom Wolfe showed in his excellent book, *The Right Stuff*, the public image of the astronauts was created by a committee of tinsmiths.

We reached the Moon amid worldwide excitement. But by that time Kennedy had been assassinated and his successors were busily murdering the space program. It was hacked to death, strangled, and smothered under piles of paperwork.

The American people allowed this murder to take place right before their eyes because they thought that the race was over and we had won it. The Russians, who saw in the early 1960s that we would zoom past them and win the Moon prize, wisely let it be known that they weren't interested in the Moon. Never had been at all. Honest. Cross their hearts.

So we won the race and relaxed, thinking it was over. Our attention was diverted by internal troubles over civil rights and the environment, by the horror and stupidity of Vietnam, by the reek of Watergate.

And the poets, the writers who had been there at the beginning when no one else would listen, who had been shunted aside when the engineers took over, the writers stood stunned and mute at the wreckage of their dream.

6

Use It or Lose It

> The Soviet Union has become the seacoast
> of the universe.
> —SERGEI KOROLEV

Think about the significance of that statement for just a moment. Think of the psychology behind the metaphor. Since the time of Peter the Great the Russians have sought a warm-water port to link their landlocked nation with the rest of the world. Korolev, the late "master designer" of the original Soviet space program, revealed an important subconscious component of the Russian emphasis on space exploration when he spoke those words. It is one of the most widely quoted statements about space within the Soviet Union, by the way, fully comparable to our own temptation to quote Neil Armstrong's words on stepping out onto the lunar surface.

We achieved something magnificent in Apollo, but we failed to follow that success with even more achievement. Instead of looking upon our work in space as an investment, we allowed ourselves to be convinced that it was a stupendously expensive stunt. We turned our backs on space. Even while we were in the full flush of Apollo's success, the program was being called a "Moondoggle" by its know-nothing opponents in Congress. We won the race and then went home, thinking that the game was over. We forgot that one race does not constitute the entire competition.

NASA's Apollo project went so well, so smoothly, so successfully, that the space program became boring. By the time the third team of astronauts was walking on the Moon, television ratings for the Apollo show were sinking. After all, once you've seen two astronauts Moonwalking you've seen them all. Right? It's all a big stunt anyway. Right? A Moondoggle. Let's get back to the important things, like the War on Poverty or the war in Vietnam.

While Richard Nixon was visiting the astronauts at their quarantine base after they returned to Earth, and smiling for the cameras, his White House staff was slashing the budget for NASA and dumping all the future plans for utilizing space for practical, long-term operations into the paper shredders.

NASA had plans for the future. Looking back on them now is like viewing the sun-bleached ruins of a beautiful ancient civilization. NASA's original "post-Apollo" scenario went something like this:

Earth orbital operations: Using components of Saturn rockets, establish temporary "workshops" in orbit close to the Earth by 1972–73 (such as the ill-fated Skylab) and permanent space stations in low earth orbit (LEO) by the 1975–76 time period. By 1980 a permanent space station would be established in geosynchronous orbit, 22,300 miles above the Equator. In such an orbit the satellite revolves around the Earth once every 24 hours, thereby remaining stationary above a fixed spot on the Equator. The LEO station would evolve into a full-fledged space base in the 1980s, where up to 100 men and women would be involved in scientific and industrial studies, including astronomy, Earth resources surveys, life sciences, space physics, and materials research and processing experiments.

Lunar operations: The very successful Apollo landings would be followed in the 1970s by landing teams of six astronauts at a time, who would stay on the Moon for several days or weeks at a time. This would gradually be extended, through the 1970s, into a permanent lunar base capable of housing at least six astronauts, who would be sent to the Moon on regular rotating schedules for weeks at a time. By 1980 there would be a 24-man station in orbit around the Moon, and by the mid-1980s a base on the lunar surface capable of housing at least four dozen astronauts and scientists who would explore the lunar surface, study astronomy, biology, selenology (the geology of the Moon), mining, and seismology, and build a planetary quarantine facility. The natural resources of the Moon would be mapped and graded for industrial development.

Planetary operations: The Viking program was originally scheduled to land on Mars in the 1972–73 period. Viking I touched down on the Martian soil July 20, 1976, four years behind the original schedule. More spacecraft were to be sent to Mars during the latter 1970s, and a Grand Tour mission of the outer planets — Jupiter, Saturn, Uranus, Neptune, and Pluto — was to be sent off by 1980. Manned landings on Mars were envisioned for the early 1980s, culminating in a semipermanent base on Mars by 1990, with 48 persons on the surface of the planet and 24 more in a space station orbiting around it.

The unmanned portions of the planetary operations program have gone fairly well, thanks in large part to the urgings and fine work of such scientists as Bruce Murray, director of the Jet Propulsion Laboratory, and Carl Sagan. Two Vikings landed on Mars in 1976. A Pioneer spacecraft is in orbit around Venus and has mapped that planet with cloud-piercing radar. Other Pioneer and Voyager spacecraft have flown past Mercury, Jupiter, and Saturn. The Grand Tour was officially vetoed by Washington, but the Voyager spacecraft that visited Jupiter and Saturn just might sneak onward to Uranus and Neptune. It would not be the first time that the scientists used a shoestring to go out on a limb.

Transportation systems: All of NASA's post-Apollo plans depended heavily on developing rocket boosters and space propulsion systems to carry out the missions. The Space Shuttle was originally planned to be in operation by the late 1970s. A space tug, useful for moving freight and personnel from one orbit to another, was planned for the latter 1970s also. By the 1980s a nuclear propulsion system and Mars excursion module were planned as the backbone of the manned Mars expeditions.

The Space Shuttle's testing program will not be completed until the mid-1980s, nearly ten years later than originally planned. (Those ten wasted years again!) No work has been done on nuclear propulsion or other advanced space propulsion systems, such as electrically powered ion and plasma engines.

During this lost decade, while our post-Apollo plans vanished in the smoke of Vietnam and civil unrest, while our Shuttle program was starved with funding cuts and stretch-outs, what have our erstwhile competitors, the Russians, been doing?

Winning the competition, that's what.

Of the more than 1000 satellites in orbit around the Earth as of this writing, about 80 percent of them were launched by the Soviet Union. In 1980, the United States orbited 17 satellites; the Soviet Union, 130.

Numbers don't mean everything, of course. America's lead in miniaturized electronics and reliability allows us to use a satellite for a much longer time than the Russians can. Still, the fact that in 1979–80 we launched a total of 34 payloads into orbit while the Soviets launched 231 shows that they are *very* active in space.

The Russians have gone to the Moon with unmanned landers that have picked up samples of lunar soil and returned them to the Soviet Union, under remote control from Earth. They have put automated roving vehicles on the lunar surface, which have traversed a few kilometers across the Moon's dusty, pockmarked plains.

And although the dozen or so spacecraft the Soviets have shot at Mars have all failed miserably, their series of Venus probes has fared much better. The *only* photographs transmitted successfully from the hellhole surface of cloud-shrouded Venus were sent by the Russians' Venera 9 and 10 landers, which touched the red-hot ground of Venus in October 1975.

These photographs, by the way, received virtually no attention from the American media. It was the first time a spacecraft had landed on the surface of another planet, and Venus' surface conditions — an atmospheric pressure high enough to crush an ordinary spacecraft and surface temperatures hot enough to melt aluminum — make Mars seem like a picnic ground by comparison. Nearly a year before Viking I's touchdown on Mars, these Russian spacecraft sent photographs back to us from Venus. No one had ever seen the surface of Venus before, because the planet is perpetually covered with thick clouds.

It was a considerable achievement, but the feat was a non-event in the American media. Why? Because by 1975 space itself was a non-event in the media, and calling attention to Russian achievements would have raised the awareness of the American people about the lack of interest in space throughout Washington — and throughout the media, as well.

Although the Russians have been quite active in exploring the Moon and the nearer planets, with mixed results, the overwhelming emphasis of their space efforts has been concentrated much closer to home.

"We believe that long-lived space stations with interchangeable crews will be humankind's main road to the universe." That policy, enunciated by no less than Leonid Ilyich Brezhnev, general secretary of the Communist Party and president of the Soviet Union, is the guiding path of the Russian space program.

In October 1969, after the Americans had smashed Russian hopes to be first on the Moon, Academician Mstislav Keldysh announced to the Soviet space planners that Russia's program to land cosmonauts on the Moon had been shelved indefinitely, in favor of a program to develop manned orbital space stations.

The result was Salyut.

The first experimental space station, Salyut 1, was launched in June 1971. Two Soyuz spacecraft, comparable to our Apollo craft and each carrying three cosmonauts, visited Salyut 1. The first team of cosmonauts docked with the space station but did not enter it. The second crew entered and stayed in orbit aboard Salyut for 24 days, a record at that time. But they were killed upon re-entry into Earth's atmosphere; apparently a hatch leaked and the air literally exploded out of their spacecraft.

Despite this shocking setback and other failures of unmanned spacecraft, the Russians pushed doggedly ahead. They redesigned the Soyuz craft to accommodate two cosmonauts, rather than three. They buried their dead and made heroes of their living cosmonauts. Today, in Russia, a cosmonaut is glorified the way American teen-agers glorify rock singers.

Up to the present at least six Salyut space stations have been placed in orbit. The Russians have used them to carry out engineering and medical tests aimed at very-long-term occupancy of orbiting stations.

The Salyuts come in two types: civilian and military. The military stations are manned by all-military crews and fly in low orbits of about 270 kilometers (roughly 170 miles), which are best suited for photographing the ground. The civilian Salyuts carry mixed crews of civilians and military officers and usually orbit a bit higher: 350 kilometers (about 220 miles).

Salyut 6, a 20-ton vehicle orbited in September 1977, has been occupied by at least two cosmonauts almost continuously since it first went up. Pairs of cosmonauts have lived and worked aboard Salyut 6 for three, four and a half, and even six months at a time. "Guest cosmonauts" from Czechoslovakia, Poland, East Germany, Hungary, Vietnam, Bulgaria, and Cuba have served aboard Salyut 6, in addition to the Russians.

The space station is resupplied by automatic "freighters" that deliver rocket propellants, air, water, food, and equipment to the Salyut. On occasion, two Soyuz spacecraft have docked with the station and four cosmonauts have come aboard.

The Russians have developed long-term life-support systems, including "space gardens" and air- and water-recovery systems, so that those precious materials can be recycled and reused.

While aboard the space station, the cosmonauts conduct studies of various manufacturing technologies performed in zero gravity, medical studies on their own reactions to prolonged weightlessness, and observations of Earth.

The Soviets are also developing their own delta-winged reusable shuttle vehicle, because they too realize that if they are to conduct even more complex operations in space, they must be able to reduce the costs of going into orbit.

Later in this book we will examine the economic and industrial importance of space manufacturing and other commercial endeavors that can be carried out in orbit. And we will take a close look at the military implications of space operations. It is quite conceivable that space will become a battleground — in fact, it seems almost inevitable that it will.

Kremlin watchers claim that part of the Soviet Union's long-range plans for political and economic domination of the West includes cornering the world's supplies of such strategic minerals as cobalt, chromium, and manganese. For years the Russians have worked hard to penetrate the Middle East and the Horn of Africa, in an apparent attempt to cut off vital petroleum supplies on which Western industry depends so heavily.

Surely the Soviet leaders understand that energy and key raw materials for industry can be obtained from space. Every indication is that they are moving calculatedly to obtain access to these resources — and perhaps to deny such access, if they can, to the West.

Take this as an earnest of the Russians' commitment to space: Their civilian space center at Tyuratam (near Baikonur) has an estimated 80 to 85 *working* launch pads. By comparison, the Kennedy Space Center at Cape Canaveral is a ghost town.

Eighty to 85 working launch pads. And that is only the civilian space center. Their military space base, at Plesetsk, has still more.

7

The Luddites vs. the Prometheans

> We are moving toward a gross economic upheaval . . . and environmental destruction on a massive, global scale . . . [There will] be several very serious, perhaps holocaustic, nuclear accidents and by the mid-1980s the United States will be ready for major political change . . . We may see the complete overthrow of the government.
>
> —KEN BOSSONG

That statement was made early in 1980, at a meeting sponsored by the Office of Technology Assessment in Washington, D.C. Ken Bossong is a bright young man with degrees in engineering and public administration. He is coordinator of the Citizens' Energy Project, a private organization dedicated to environmental protection and solar energy. He is the author of five books and more than a hundred articles and reports on energy policy and other issues.

He is a Luddite.

History has not been kind to Ned Ludd, the unwitting founder of the Luddite movement of the early nineteenth century.

Webster's *New World Dictionary* describes Ludd as feeble-minded. The *Encyclopaedia Britannica* says he was probably mythical.

The Luddites were very real, however. They were English craftsmen who tried to stop the young Industrial Revolution by destroying the textile machinery that was taking away their jobs. Starting in 1811, the Luddites rioted, wrecked factories, and even killed at least one employer — who had ordered his guards to shoot at a band of rioting workmen. After five years of violence, the British government took harsh steps to curb the Luddites, hanging dozens and transporting others to prison colonies in far-off Australia. That broke the back of the movement, but not of the underlying causes for the movement. Slowly, painfully, the original Luddite violence metamorphosed into political and legal actions. The labor movement grew out of the ashes of the Luddites' terror. Marxism arose in reaction to capitalist exploitation of workers. The Labour Party in Britain, and socialist governments elsewhere in the world, evolved out of that early resistance against machinery.

Today the descendants of those displaced craftsmen live in greater comfort and wealth than their embattled forebears could have imagined in their wildest fantasies. Not because the employers and factory owners suddenly turned beneficent. Not because the labor movements have eliminated human greed and selfishness. But because the machines — the machines that the Luddites feared and tried to destroy — have generated enough wealth to give common laborers houses of their own, plentiful food, excellent medical care, education for their children, personally owned automobiles, television sets, refrigerators, stereos, and all the other accouterments of modern life.

It is commonplace to denigrate the gadgets with which we surround ourselves. But I am the son of a laboring man and a descendant of Italian peasants. I grew up in a row house that could most kindly be described as "modest." Yet I took it for granted that I could listen to the finest musical artists with the click of a switch. I could read the works of the greatest minds of civilization simply by walking down the street to the public library. I could admire masterpieces of painting and sculpture by riding a trolley to the Philadelphia Museum of Art. And by stepping into the Fels Planetarium of the Franklin Institute, I could begin to see the wonder of the wide, starry universe.

Yet today we still have Luddites among us: people who distrust the machines that we use to create wealth for ourselves. These new Luddites are most conspicuous in their resistance to high technology: computers and automated machinery, nuclear reactors, high-voltage power lines, airports, chemical products like fertilizers and food additives.

The new Luddite movement is neither unified nor preplanned. It consists of elements from environmentalists, racial and ethnic minorities, consumer protection advocates, and frustrated suburbanites worried about the quality of their lives.

Much of what these groups are trying to achieve is necessary and good. They are among those who have been fighting carcinogenic additives in our foods, unsafe factories, unsafe cars, and unhealthful environments.

But all social movements have a dynamic of their own. They must grow or die. They must find new food to grow on. For the Luddite movement, with its antitechnology appetite, high-technology efforts such as the space program and nuclear power are like feasts set before their eyes.

The original environmental and consumer protection movements had to fight enormous indifference and enmity from both government and industry. Inevitably, this produced a negative bias, a knee-jerk reflex against anything touched by "the enemy." To today's Luddites, *any* program involving high technology is under immediate suspicion. In their view, technology is either dangerous or evil or both, and must be stopped before it's too late.

Their automatic response is negative.

Their favorite word is "no."

Their tactics are delay, harassment, and the prevention of movement in any direction.

Inevitably, the Luddites throw up massive roadblocks to stop the growth of any new technology. They point to *The Limits of Growth* to prove that anything new will lead to catastrophe. Without realizing it, they aided the Nixon and later budget-cutters in slashing the space program almost to death.

And, just as inevitably, they begin to react to symbols instead of realities. Automation is evil because it throws people out of work, the Luddites believe, so they are against computers. All computers, even the ones used in hospitals to help save lives, even the ones used in social welfare agencies to help keep records straight. If you tell them that more Americans have jobs making, selling, and servicing computers than have been put out of work by automation, they either refuse to believe you or simply make a mental sidestep and ignore that part of the situation.

Nuclear power plants are a favorite target of the Luddites, representing to them a combination of Big Business, Big Government, and Big Military. In truth, there is much improvement to be made in the design, operation, and maintenance of nuclear power plants; and Luddite opposition to "nukes" has helped to balance an equally intransigent position on the side of the nuclear power industry. But demands to shut down all nuclear reactors immediately simply do no one any good. Yet if the Luddites had their way, the nukes would be shut down, and many areas of the nation would go dark.

The Luddites march, they protest, they lobby in Washington, they write academic papers and emotional books. They try to pull the plug on the future, to stop the clock where it is or perhaps even wind it back to a calmer, safer, happier time. Philosophically, they point to Jefferson's view of a decentralized America, populated with hardworking, happy yeomen farmers.

Opposing the Luddite point of view stands a group of people

who fear neither technology nor the future. Instead, they rush
forward and try to build tomorrow. They are called Prometheans, after the demigod of Greek mythology who gave the
gift of fire to humankind.

Every human culture, throughout history, has created a
Prometheus myth, a legend that goes back to the very beginnings of human consciousness. In this legend, the first humans
are poor, weak, starving, freezing creatures, little better than
the animals of the forest. A godling — Prometheus, Loki, Coy-
ote, whoever — takes pity on the miserable humans and brings
down from the heavens the gift of fire.

The other gods are furious, because they fear that with fire
the humans will exceed the gods themselves in power. So they
punish the gift-giver, eternally.

And, sure enough, with fire the human race does indeed
become master of the world.

The myth is fantastic in detail, but absolutely truthful in
spirit. Fire is the symbol of technology. Technology is our way
of dealing with the world around us, our path to survival.

As the English biologist J. B. S. Haldane observed, "The
chemical or physical inventor is always a Prometheus. There is
no great invention, from fire to flying, that has not been hailed
as an insult to some god."

*

I count myself among the Prometheans. Perhaps I am preju-
diced. I started my writing career as a newspaper reporter in
the late 1940s. In those days, every summer the nation's newspapers carried a long, ugly, running story about polio. It was like covering the baseball season; all summer long we ran box scores every
day on the number of children who had been killed by polio,
the number placed in iron lungs, the number crippled for life.

Then one springtime in the mid-1950s we carried one single
story. Lots of human interest. Plenty of wonderful photographs.
Children were being inoculated with the Salk vaccine. Good

front-page stuff: a kid screaming bloody murder as a doctor jabbed a needle into his arm and his anxious mother smiled bravely in the background.

That was a dull summer, as far as polio went. There has never been another summer when any newspaper in the land has had to carry a running account of polio's ravages.

That's why I am a Promethean. Science is knowledge. Technology is tools. Knowledge and tools are not to be shunned; they are to be *used*.

*

In the labyrinthine twists that mythology often takes we have substituted punishment of the gift-giving god for what really happened: punishment of the gift-receiving humans. For by accepting fire, by accepting technology, we have accepted a not-unmixed blessing. As the Luddites correctly point out, technology leads to pollution, oppressive social structures, loss of individual dignity. Better to live a "natural" life, they say, than to be dehumanized by runaway technology.

Yet technology is as natural a part of human life as walking. Before *Homo sapiens* existed, our apelike hominid ancestors were making and using tools. Their chipped pebbles are the earliest artifacts on this planet. The primitives who received the gift of fire from the heavens (most likely a lightning stroke that set a bush aflame) were already technologists of a sort. Technology — tool-making and tool-using — is as natural to us as flight is to a bird. A human being without technology is not a noble savage, but a dead naked ape.

But suppose we could turn the clock back, as the Luddites long to do. Suppose we could return to a calmer, earlier time: say, the America of 1905, that wonderful Middle America of gazebos and hobble skirts that we picture with such nostalgia at Disneyland.

Who would die?

Global population was less than two billion in 1905; the population of the US was a little more than one-third of what it is today. How do you get rid of those extra people? We could not support them, not with the technology of 1905. Farmers with horse-drawn plows and family doctors in horse-drawn buggies could neither feed nor medicate four and a half billion people. More than half the human population of Earth would have to die. Who would it be?

The ghetto dwellers of every major city in the world? The villagers and herdsmen of India, Ethiopia, Guyana, Mali, Indonesia? The teeming millions of China and Japan? Or the suburbanites of America, dependent on their automobiles and tranquilizers?

If we turned our backs on technology, if we closed the factories and shut down the power plants and went back to organic farming and returned to "nature," billions of people would die in a very few months. Assuming that they would be content to die peacefully, of starvation and disease, without riots and revolutions and wars.

The Luddites who campaign for a simpler world tacitly assume that they — white, affluent, educated — will be the survivors and beneficiaries of the "simpler" world they desire. A leaky assumption, at best.

Obviously, no one can stop the clock that abruptly, and not even the most insensitive Luddite believes that the solution to the world's problems is to kill off two or three billion human beings.

From a practical point of view, how can we stop the clock slowly, gradually wind things down? Gradually cut down industrial production? Enforce laws that will control family size? Use the power of the state to lower birthrates, set up limits to growth, put ceilings on farm output? As we have already seen, to enforce limits on population growth, industrial production, and farm output on a global basis would require a worldwide dictatorship of such crushing power that it is too bitter even to contemplate. George Orwell's darkest vision of 1984 would come alive everywhere, for everyone:

If you want a picture of the future, imagine a boot stamping on a human face — forever.

Imposing limits from above means curtailing human freedom. It is the way of the unconscious fanatic who is going to tell *you* what is best for you whether you like it or not.

*

The Promethean way looks toward the future and toward our ability to use technology to shape our destiny. As Lewis Branscomb, vice president for research at IBM, put it:

Technology has brought us changes, most of which we should welcome, rather than reject. Wealth is the least important of these changes. Of greater importance is change itself. Those young humanists who think themselves revolutionaries are nothing compared to technology.

The Luddites see technology as part of the problem. The Prometheans see it as part of the solution. It was technology, the Prometheans point out, that freed the slaves. In ancient Greece a visitor observed that the vaunted Athenian democracy was built on the sweat of slaves. Aristotle reportedly retorted, "When the looms spin by themselves, we'll have no need for slaves." The steam technology that ushered in the Industrial Revolution killed off slavery. Not the religious moralists. Not the revolutionary firebrands. Not the kindly plantation owners or the hardheaded businessmen or the working free men. It was the machines. Once the looms spun by themselves, once steam power became cheaper than human muscle power, slavery withered and died.

Today we are in the throes of a Second Industrial Revolution, where electronic computers are taking over tasks that humans once performed. The Luddites distrust this revolution as much as their forebears hated the Spinning Jenny. But the Prometheans welcome it. They see it as an aid in solving the problems that face us. Automated machinery will make us all richer by increasing the productivity of the worker. Steam power replaced physical drudgery; why shouldn't electronics replace mental drudgery?

The Promethean bargain is to accept technology, embrace it, use it, on the assumption that the cost will be worth the gain. At first we accepted technology — fire, this gift from the heavens — without a thought about the cost. A shivering Stone Age hunter does not worry about environmental pollution when he strikes up his fire. A sweating Bronze Age peasant has no thought for altering the climate when he chops down trees to make room for farmland.

As the village blacksmith evolved into the smoke-belching "dark, Satanic mills" of the Industrial Revolution, people cared about jobs and profits, food and money, not about ecology. We accepted technology on the simplest terms possible. We asked our technology to feed us, to keep us warm, to protect us from our enemies — *regardless of the consequences.*

Our technology granted us those wishes so well that the consequences now threaten to overwhelm us. We have fed ourselves to the point of overpopulating the planet, warmed ourselves so thoroughly that we may soon melt the polar icecaps, and built weapons of protection that could destroy the world. The Promethean answer to this dilemma is not to retreat into the past, but to press forward and create a better tomorrow with a *second-generation technology* that can feed us, warm us, and protect us without undue damage to the environment.

Is this possible? Can we create a second-generation technology that can accommodate five or six or seven billion people without totally wrecking our world? The Prometheans believe that it can be done and that space technology is the vital element in accomplishing this crucial task. We must utilize the resources of space to solve the problems of Earth.

8

The Rhetoric of Energy

> Now is the time for the United States to
> come to terms with the realities of the en-
> ergy problem, not with romanticisms, but
> with pragmatism and reason. And not out of
> altruism, but for pressing reasons of self-
> interest.
>
> —ROBERT STOBAUGH AND
> DANIEL YERGIN

Nowhere is the difference between the Prometheans and the
Luddites more sharply drawn than in the clamorous battle over
the energy crisis. With all the voices raised about various energy
matters it is difficult to tell who wants to do what, or where our
politics are heading. But beneath all the public rhetoric and
posturing lies a sharp division of social values, and beneath *that*
lies the central political-economic struggle of this century.

There are three — count 'em, three — main camps in the
energy battle: the Establishment, the Luddites, and the Prometheans.

The Establishment consists of a mix of major corporations and
federal agencies. Between them, they make most of the business and governmental decisions for the US. To outsiders it

often seems as though government and business are locked in a struggle against each other, particularly over price regulations for such fuels as petroleum and natural gas. But these debates about price controls and deregulation are something like a typical science fiction film: plenty of special-effects pyrotechnics, little of lasting significance or interest. Oil and natural gas are being used up at a fearsome rate. Resources that took nature hundreds of millions of years to lay down have been sucked out of the ground and burned in little more than a century. The loud and showy debates over price regulations are merely squabbles over how much to charge for the last few drops of oil and natural gas.

The real question is: How do we replace petroleum and natural gas as their availability dwindles and their price (no matter how the debate goes) escalates?

The Establishment answer, as you might expect, is conservatively based on known resources and technologies. The Establishment wants to use coal and uranium as our main energy sources for the near term and then move on to renewable sources like solar energy and, in the dimly perceived far future, thermonuclear fusion. Conservation — simply cutting down on energy usage — is a critically important part of the Establishment plan.

Coal can be used directly as a fuel, or converted into synthetic liquid hydrocarbon fuels for transportation uses. Since coal is dirty, dangerous, carcinogenic, and causes acid rain, the Establishment plan includes efforts to "scrub" the stack gases of sulfurous pollutants and soot wherever coal is burned directly, as in electric utility power stations.

Nuclear energy presently provides as much as 12 percent of the nation's electricity, and despite accidents such as the Three Mile Island fiasco, the Establishment calls for building more (and better) nuclear power plants. In the long term, breeder reactors are called for, to stretch our limited supplies of uranium so that they will last indefinitely.

Conservation has a negative connotation, so there is an Establishment effort to rename it *energy efficiency*. Whatever you call it, it makes good sense. Better insulation for houses, fuel-efficient automobiles, more energy efficiency in factories — a vigorous effort along these lines could cut down energy usage by almost half. In thirty to forty years.

There are flaws in the Establishment plan. Coal and uranium can have disastrous health and environmental impacts. Converting coal into synfuels runs smack into another scarcity: water. It takes enormous amounts of water to run a coal conversion plant, and the areas of the US where coal is most abundant happen to be precisely those areas of the West where freshwater supplies are most limited.

Burning coal, or any hydrocarbon fuel, can have very serious long-term environmental effect, as well. Even the cleanest stack gases pump megatonnages of carbon dioxide into the atmosphere daily. Carbon dioxide traps solar heat, creating a "greenhouse effect" in the atmosphere. Relying on fossil fuels for our energy needs inevitably carries with it the threat of warming up the Earth's atmosphere to the point where global climate is affected. Long before the icecaps melt, the growing seasons of the world's granaries will change. What will happen to the American Midwest, the Ukraine, the rice paddies of Asia, the wheatfields of Manitoba and Argentina? No one can make a long-range forecast of any accuracy at all.

Coal-burning also causes acid rain, which destroys forests and turns lakes into sterile ponds of sludge. Although sulfur emissions from coal can be curbed, a quantum jump in coal usage will inevitably lay waste to more farmlands, forests, and water resources.

Conservation, or energy efficiency, takes a long time to make a significant contribution. In strict commercial terms, we would have to restructure our marketing priorities and even our bookkeeping to provide incentives to individuals and businessmen to move toward more efficient use of energy. At present, for example, the federal subsidy to a private homeowner for insulating his or her house runs up to a limit of $300. Despite all the brave talk in Washington, that is no incentive at all for most homeowners.

And despite the American automobile industry's newfound faith in small economy cars (forced on the dinosaurs of Detroit by public preference for foreign imports) the advertisements for cars still emphasize the auto as a sex aid, not a mode of transportation. The advertised mileage ratings are pitifully low, and the cars don't even live up to those numbers after they have been on the road for a while.

Moreover, Detroit has cleverly levered Congress into relaxing environmental requirements on automobile pollution effluents in exchange for those minor increments in mileage ratings.

The Establishment plan, then, is aimed at weaning us away from petroleum and natural gas as slowly as possible. The proposed alternatives satisfy almost no one, which is why the Establishment plan is under attack almost constantly from every side.

The Luddites' energy policy is against "big" technology and in favor of "small" technology. Amory Lovins's "soft energy paths" is the Luddite gospel: Big utility power plants are "hard" technology and therefore evil; solar heating for individual homes is "soft" technology and therefore blessed.

To their credit, the Luddites see clearly that energy decisions are also environmental decisions. They worry about the ecological threats stemming from factories, power stations, and automobiles.

They see the main problem as the inevitable outcome of human greed and big, inhuman, centralized institutions such as electric utilities, multinational corporations, and government bureaucracies. As much as they fear the ecological damage done by burning fossil fuels, the Luddites fear equally the enormous power of mammoth centralized institutions of Big Business and Big Government.

Their solution is to turn to solar energy, which they see as clean, cheap, abundant, and decentralized. Remember that word: *decentralized.* It is the key to most of the rhetoric about energy that rings through the land.

The Luddite argument for decentralized solar energy is based partly on ecological and partly on political grounds — with strong moral overtones, as well.

Big Business and Big Government are the major culprits, as the Luddites see it. Big, centralized, "hard" technologies like nuclear power are the tools of oppression. The victims are the people and the environment.

Although energy shortages and environmental degradations exist in the Soviet Union and other socialist nations, the Luddites point with special anger at America and Americans as short-sighted, overweight, selfish energy gobblers.

The United States uses nearly 30 percent of all the world's fossil fuels. In fact, almost 30 percent of all the energy consumed in the world is consumed by the US. On a per-capita basis, each American consumes enough energy per year to equal slightly more than eight metric tons' worth of petroleum. This is considerably more than other people use. The average Russian consumes four metric tons' worth of oil per year, a bit less than half the American figure. Consumption for West Europeans is 3.27 tons; it is 3.06 for Japanese, 0.92 for Latin Americans, and 0.41 for Africans.

We are energy hogs, say the Luddites. We take more than our fair share.

But this assumes that there is only so much energy to be had, that our supplies are strictly limited, and that for an American to consume more energy than a Ugandan is inherently wrong. Instead of attempting to find more energy resources, this argument falls into the Flat-Earth Fallacy. It presupposes that we must continually slice a finite energy pie into thinner and ever-thinner portions, so that everyone can have an equal share.

Yet if our supplies are truly finite and our numbers continue to grow without check, sooner or later the shares get so thin that we all starve to death. Equally.

To get us off the oil habit and away from fossil fuels altogether — indeed, to get us out of the grasp of the big corporations and the government — the Luddites want to turn to renewable energy

sources: energy resources that ultimately depend on the Sun.
Solar energy comes in a variety of forms. There is *solar heating,* where sunlight is used directly or indirectly to provide heat
and/or hot water for a building. There is the *solarvoltaic cell,*
solar cells such as those used in spacecraft, which convert sunlight into electricity for home or other uses. There is *biomass,*
in which trees or other vegetation (or even garbage) that have
trapped solar energy through photosynthesis are burned for
heat or used as a heat source to generate electricity. These are
all "soft" energy technologies.

Hydroelectricity, wind power, and various schemes to extract
energy from temperature differences in the sea all depend on
solar energy. The Sun evaporates seawater, which then falls as
rain to fill the lakes that provide energy for hydroelectric dams.
Sunlight heats the atmosphere and oceans to drive the winds
and cause ocean thermal gradients.

Wind power is considered a "soft" technology because it is
decentralized; you can build or buy your own windmill and
unhook yourself from the utility company's power line, in principle. Hydroelectric dams and ocean-thermal machines require
major centralized organizations and are therefore considered
"hard" energy technologies.

Sunlight is clean, dependable (within the vagaries of the
weather), abundant, and inexhaustible. The Sun has been shining steadily for at least four billion years, and astrophysicists
calculate that it has another eight or ten billion years of uninterrupted radiance ahead of it.

To be strictly accurate, fossil fuels are also a form of trapped
solar energy. Coal, oil, and natural gas are the remains of living
organisms that perished geological ages ago and have been
fossilized. Burning these fuels releases the energy that they
accumulated from sunlight all those eons earlier. Buckminster
Fuller, the grand old man of futurism, has said many times that
it's time to stop burning fossil fuels and turn on this planet's
"main engines," meaning solar energy in all its many manifestations.

Can we "go solar," as the Luddites insist we must? Can we stop using fossil fuels and uranium altogether?

Each year, the United States consumes something like 18 trillion kilowatt-hours of energy (18 × 10¹² kWh). That is our total energy usage, including everything from home heaters to the blast furnaces of steel mills, from campfires to the Astrodome's air conditioning, from mopeds to jumbo jets.

Eighteen trillion kilowatt-hours per year. That's equivalent to the energy output of exploding a thousand one-megaton hydrogen bombs. Huge as that amount of energy is, solar energy actually provides more. Much more.

The Sun bathes our nation in roughly 740 times more energy than we consume, directly in the form of sunlight. The winds blowing across our land contain five times more than that 18 trillion kilowatt-hours per year. The thermal gradients in the oceans, if tapped for energy, could yield six times the amount we consume annually.

To put it in terms that are easier to grasp, sunlight delivers almost 70 watts per square foot of ground at the latitudes of the contiguous forty-eight States. Alaska, being farther north, where the winter days are short and the Sun is at a low angle even during the summer, gets considerably less energy from the Sun. Hawaii, close to the Equator, gets considerably more.

Seventy watts per square foot means almost a kilowatt (one thousand watts) per square yard. How many square yards of roof surface to your home?

It means more than three megawatts (million watts) per acre. How big is your farm?

Nobody can use that energy with 100 percent efficiency, of course. But it's free, it's abundant, and it falleth out of the sky on the just and unjust alike, every day. Solar energy could provide a low-pollution energy source for us all, if we had the means to utilize it. As a character in a Robert A. Heinlein novel said: "It's raining soup; grab yourself a bucket."

Equally important to the Luddites, remember, is that ground-based solar energy systems must, by their very nature, be decentralized. There is a bitter joke told by solar-energy enthusiasts: "When the power companies figure out how to get a sunbeam through an electric meter, then we'll get solar energy." The reality behind the joke is that the "power companies" — the electric utilities, the oil corporations — know full well that decentralized solar energy on a major scale could wreck their businesses. They have no desire to help promote solar, wind, or any other form of energy technology that will hurt them.

The smart companies are acquiring or developing divisions that sell solar and other "soft" energy technologies, while at the same time heavily promoting the continued use of fossil fuels. Take a look at oil company advertisements in any major magazine or on television. Whenever they mention solar energy, it is in the context of something that is desirable, but so far off in the future that you'd better not depend on it for a long, long time to come.

The less-smart companies, such as most electric utilities, are resisting the move toward solar and "soft" energy paths. Con Edison, in New York City, even got the utilities commission to pass a regulation penalizing anyone who set up a windmill to generate electricity. The ruling was swiftly overturned once consumer groups learned of it, and now there is a federal law prohibiting such actions.

Nuclear power, of course, is the epitome of all the Luddites fear: Big Business and Big Government, together with the Military, uniting to force a dangerous, expensive, potentially disastrous "hard" energy technology down the throats of the people. That is what is really behind the cry of "No more nukes!"

The Establishment sees nuclear power as an absolute necessity. Less nuclear means more oil today and, in the near future, more coal — which is far more dangerous than uranium to the environment and to human health. The Luddite answer is to "go solar." But, their critics counter, the kind of solar energy technology the Luddites want just isn't here. Soft technology though it may be, it simply does not exist yet, not on the scale needed to turn the US into a decentralized Solar Democracy. You can build a solar hot-water system for yourself, with a little ingenuity and some carpentry and plumbing. But it will be years before the cost of that system is amortized by your lower fuel bills. Most people will not bother; they will simply keep on paying higher and higher fuel bills. They'll grumble, but the economic return for "going solar" is too slow to pay off.

I investigated the possibilities of building a solar house in Connecticut, where my wife and I reside. After a year of detailed professional design work, the conclusions I came to were:

1. No one could say for certain how much of the basic heating load for the house the solar equipment would provide; the technology is too new to have accurate performance information available.

2. Whatever heating the solar equipment did not provide I would have to purchase, in either electricity or heating fuel.

3. There were no measures of reliability of the solar equipment, other than, "If anything goes wrong, we'll come out and fix it."

To put it bluntly, I can heat a home in Connecticut with electricity provided by nuclear power much more reliably and cheaply that I can with solar energy.

Aha! say the Luddites. That's because Big Business and Big Government don't want solar energy, don't want to be broken up by millions of self-sufficient homeowners making their own heat and electricity. So they won't support the research and development needed to make solar technology practical, reliable, and easily available in the marketplace.

The Prometheans look upon all this with a tolerant — nay, skeptical — smile. Sure, they maintain, solar energy is fine. As far as it goes. And yes, the Establishment plan is far too concerned with immediate political urgencies to provide a solid approach for the long-term future. As for the decentralized-vs.-centralized argument, the Prometheans feel that there is a need for both. And plenty of room for both, as well.

Decentralized solar energy systems are wonderful for decentralized uses, such as individual homes. But transportation systems, from private automobiles to farm machinery, will require some sort of liquid fuel to replace the petroleum they now burn, unless and until the electric car (as well as tractor, truck, harvester, bus, airplane, and so on) comes along. A gossamer-winged airplane has successfully flown, powered by solarvoltaic cells. But running the family auto on sunlight seems a long way off.

There will continue to be a need for centralized energy sources: fuel distribution systems for transportation; power stations that can deliver megawattages of intensive energy, whether it is in the form of electricity or heat. Factories, steel mills, and cities cannot run on low-grade, spread-out, decentralized solar or wind systems.

If solar energy is a soup that falls freely out of the sky, the Protheans say, it's a thin soup: perfectly adequate for individual homes and small, low-energy consumers. But big cities and factories and aluminum smelting plants need high-grade, intensive energy sources, a steaming thick stew rather than clear broth.

Solar Democracy is fine, but it will not run our major cities, nor our big factories.

Nuclear power and coal are both too environmentally dirty to use one moment longer than we have to. The Protheans insist that we must look beyond them to newer energy technologies and to a future where a blend of soft and hard energy approaches not only solves our existing energy worries, but produces an energy-rich world.

Foremost among the Promethean hopes is to use the knowledge we have already gained in our space program to deliver these new energy technologies. From the work already done on rocket engines and fuels, we have learned how to produce and use hydrogen as a fuel. Clean and abundant as water, hydrogen could replace fossil fuels everywhere. From the research on advanced heat-shield materials and the hypersonic flow of incandescently hot re-entry gases, we have learned how to build new types of electric power generators that can deliver multimegawatts of electricity cleanly, even when burning high-sulfur coal.

We can develop nuclear fusion, the energy source of the stars themselves. We can build solar power satellites, which tap the Sun's energy in space and beam it back to Earth. We can use the engineering and management techniques gained from complex programs like Apollo to make existing nuclear power plants safer. We can boost radioactive nuclear wastes off the planet.

But the Promethean goals horrify the Luddites. Each of these energy options appears to them to be dangerous, hopelessly complex, inevitably unreliable, and — worst of all — requiring large, centralized organizations to carry them out. Here is the basic power struggle of our century: the centralizers vs. the decentralizers; the big vs. the little; the corporations vs. the people.

Underlying all the rhetoric about energy is this fundamental controversy. Behind the cries of "No more nukes!" and the three-way argument among the Establishment, the Luddites, and the Prometheans is this continuing struggle between the growing power of the corporations and the fears of the private citizens.

From the Civil War onward, the motivating force behind American history can be viewed in the light of this struggle. The rise of the powerful Rockefellers, Morgans, Carnegies, and the so-called robber barons. The antitrust laws of the Theodore Roosevelt and Wilson eras. The basic philosophical and demographic differences between the Republican and Democratic parties. All are a reflection of this fundamental struggle.

Today, as political party lines blur and demographic loyalties shift, the battle lines become indistinct, harder to see. But the battle shines through in vivid clarity in the rhetoric over energy, decentralization, and the potential use of space to solve our energy problems. Will we live in a Corporate America, where the decisions made in the board rooms of a handful of giant companies will outweigh those made in Congress and the White House? Many feel this is already the case, and that our government's decisions are little more than rubber-stamp approvals of corporate demands. "The business of America is business," said President Calvin Coolidge in the 1920s. Sixty years later, many Americans fear that business runs the government, here and abroad.

Corporate America or Solar Democracy? That is the real dividing line behind the rhetoric of energy. And, like a tiny strip of land caught between two warring kingdoms, the space program has become a pawn in the power struggle between these opposing forces. Neither side can tell for certain if the space program will help or hurt its cause. It seems quite clear that we can use space technology to provide abundant energy for all; later chapters of this book will show that in detail.

But abundant energy is only one of the things that a vigorous space program can provide for us all. The first, most obvious benefit. Beyond that comes a total transformation of human existence, as we develop and utilize the raw materials, the natural resources that await us in space. We know enough about the rest of the Solar System now to realize that we can bring so much wealth back to Earth from space that it will change the world's economic and social conditions just as radically as the discovery of the New World changed European life, six hundred years ago.

Will such changes help the corporations or the people? Will the riches brought back from space benefit the cause of centralization or decentralization? My suspicion is that it will help us all, and that the wealth awaiting us in space is so vast that it will overwhelm today's arguments between Big and Little, just as the discovery of America silenced the wars between Christianity and Islam.

9

The Politics of Scarcity

> Give us the tools, and we will finish the job.
> —WINSTON CHURCHILL

One of the most enduring themes of literature is the tale of the beggar who is really a prince. Scientists and technologists often appear to the public as beggars, pleading for funds, asking for our attention, our understanding, our help, promising us wonders if we will only give them a few coins.

The strange truth is that, most often, when we give them the funds they seek they produce exactly what they promised. But is it what we really wanted, in the first place? According to author Kirkpatrick Sale:

> We have tried the future — and it doesn't work.
>
> For 30 years now, this nation has been on a relentless and expensive high-technology binge, forging for itself the machines and systems that are supposed to underpin — and presage — our 21st-century lives. The only trouble is that all this high technology not only doesn't seem to be solving our problems, it actually looks to be compounding them.

Sale then goes on to decry Valium, synthetic fuels, antibiotics, supersonic transports, birth control IUDs, the Law Enforcement Assistance Administration, and the helicopters used in the aborted rescue mission in Iran in April 1980. "Solutions," he concludes, "are very much like problems: They are rooted in people, not technology."

That sounds almost profound, but what does it mean? People without technology have no means of dealing with problems. We are a technological species, like it or not, as dependent on our tools as a gazelle is on its swiftness or a tiger on its claws.

The problems are rooted in people. The world's human population is growing faster than our ability to feed it. And by "feed," I mean not merely to provide food, but clothing, housing, education, individual dignity, and freedom. The most obvious manifestation of this fundamental problem, to Americans, is the energy crisis. But to other peoples, shortages of food, shortages of jobs, shortages of *hope* are the immediate realities. To this the Luddites reply, we are entering an era of scarcity. We will all have to learn how to live with less. And all around the world, population increases and resources dwindle.

Day by day the trap becomes tighter and tighter. Where once you could roam freely through national parks like Yellowstone and Yosemite, enjoying the wilderness, today you must reserve your time and space and elbow your way through the crowds. Where once an employee of demonstrated incompetence could be fired, today he is kept on, and two more employees as indifferent to work as he are added to the payroll. Television stations in "the land of the free" lock their doors and hire armed guards to protect themselves against dissidents. Fanatics set off bombs in public buildings for no other reason than to draw attention to themselves. These are minor symptoms of the politics of scarcity, incremental tightenings of the trap we are in. Starvation and war are major symptoms, and both are appearing all around the world.

To solve these problems, to escape this trap, to break free of the politics of scarcity, we must move in new directions. But the Luddite reaction is always *No.* No, to anything new. No, to improved technology. No, to anything except a retreat into the past.

The Prometheans draw strength from the wisdom of evolution: Change or die. They recognize that the beggar is a prince who holds within his mind the power to break us free of the tightening trap we have fashioned for ourselves.

To solve the problems of Earth we must look beyond the Earth. Instead of resigning ourselves to the politics of scarcity, we must create the politics of plenty. To break the trap of growing population and dwindling resources we must enlarge our supply of resources.

Space is not an escapist dream. It is the oil field, the silver mine, the breadbasket for the world. Space is not a region to be "conquered" or colonized. We will not, cannot, solve our problems by shipping "excess" population off-planet. But space *is* a region where a few men and women can tap the resources that our Earthbound billions so desperately need. To accomplish this, we must think and act in terms of decades, not merely years. We must plan and work for long-range results. We are not accustomed to thinking and planning very far ahead. When the Arab Oil Embargo cut off petroleum imports from the Middle East in 1973, Prometheans immediately saw that we must develop alternate energy technologies like solar, wind, hydrogen fuels, power satellites, coal gasification, and others.

"But that would take ten years!" cried the Establishment and the Luddites, in chorus. It will still take ten years to develop those technologies. If we had started in 1973, we would be almost home by now.

Prometheans are often taunted with the question, "If you fellows could get to the Moon, why can't you . . ." The rest of the question is filled in with whatever problem is bothering the speaker at that moment. The answer is, almost unfailingly, *we can.*

Whatever problem you can imagine, if it has a technological component to it, chances are that the technology to solve the problem already exists. From energy to education, from hunger to transportation, the problems can be solved. We have the knowledge. We can build the tools. We have the wisdom to develop second-generation technologies that can solve our problems without degrading the environment.

We can escape the trap. We can break free of the politics of scarcity. But to do so, we need courage and foresight. We must look beyond the immediate passions of the day, beyond the next paycheck or next vacation or next election campaign. We must look even beyond the century-long struggle between the corporations and the people. We must take the long-range view and build solidly for the future. The alternative is a worsening of our problems, and the irrevocable closing of the trap of poverty and despair. The politics of scarcity is a self-fulfilling prophecy; a prophecy we dare not allow to come true.

II

THE OPPORTUNITY

> Sir, we are not weak if we make use of those
> gifts which the god of nature has given us.
> —PATRICK HENRY

10

Future Two

The year is A.D. 2000.

The biggest hit in the pop music field is a Golden Oldie, "River of Smoke," originally recorded more than half a century ago by a group called Fred Waring and His Pennsylvanians. It tells the story of a man who has a good job in the steel mill, and he's happy to see the smoke pouring up from the stacks every morning, because that smoke means he's "workin' all day, makin' my pay, savin' it up for that swell little gal someday, that gal that I'll marry someday."

Jobs are plentiful, although good jobs with the big corporations, where a man or woman can have clean fingernails at the end of a shift — those are scarce. Kids start competing for the notice of the corporation headhunters while they're still in junior high.

National Coal Week is coming up, and maybe that's why the big public relations companies have put so much push behind "River of Smoke." They like to impress the corporations, too. Get a client's nose out of joint and the next thing you know you'll be waiting in line for a job in the mill. Or the mines.

Coal saved America from going under. Sure, some of the medical people give out scare stories about the rising rates of cancer and lung disease, and the few eco-freaks that haven't been rounded up still write their underground tracts about acid rain destroying the Great Lakes and the croplands of the Midwest. But who pays attention to those nuts? Nobody who wants to work regularly.

There's still oil to be had, if your company can afford it. And if your corporate executives know how to speak Arabic.

A Presidential election is coming up in November and the corporations have promised to pick candidates that will really be exciting. One of them will even be allowed to make speeches about solar energy, they claim. He'll be the loser, of course.

11

The Danger of Either/Or

> I do not shrink from this responsibility
> —I welcome it.
> —JOHN F. KENNEDY

During the height of the Vietnam War the comedian Pat Paulsen commented, "Kids today say, 'Make love, not war.' Hell, during World War II we did both."

One of the problems with our energy rhetoric, with our political thinking in general, is that we tend to see the world in terms of either/or. Either we continue to import Middle Eastern oil or we freeze in the dark. Either we increase our military budget or we cave in to the Russians. Either we stop industrial growth or we choke on our own pollution. Either we fund space exploration or we fund social welfare programs. Either you're for me or against me.

We have fallen into an either/or attitude, even, in the way we think about our relationship with our natural environment. We tend to think of nature as something apart from us, or at least as something we've grown distant from, and we think of human civilization as something distinct and quite separate from nature. Often we see these two as antagonists; certainly we have been trained to think of civilization as a threat to the natural environment.

A rancher cutting back sagebrush that is invading his pasturage, a housing project swallowing up open land, the Alaska pipeline traversing tundra and muskeg despite environmentalists' protests, a suburbanite battling crabgrass in a lawn — there are plenty of examples of the *"either* civilization *or* nature"* competition all around us.

Yet most of what we consider to be "nature" is actually humanmade, as the eminent microbiologist René Dubos has pointed out. A Pulitzer Prize–winning author as well as a scientist, Dubos makes a clear distinction between wilderness, nature, and civilization. In *The Wooing of Earth* (Scribner's, 1980) Dubos says, "Some of the landscapes that we most admire are the products of environmental degradation." This is no technological philistine speaking, but a man who is intimately aware of the dangers of either/or thinking, of seeing the world through rigid doctrinaire blinkers.

Dubos cites as an example the bleached-bone landscapes of Greece, much admired by tourists and the Greeks themselves, which are the result of deforestation in ancient times. The famous Hymettus honey of Attica arose after the forests had been cut away and, according to Plato, "the mountains only afforded sustenance to bees."

Before human intervention there was only wilderness. But hardly a square meter of Earth's land surface has not been altered by human activity. From Neolithic hunters who roamed across continents and slaughtered entire herds of bison and mammoth to Bronze Age farmers who felled forests and remade the shape of the land, humans have left very little of this planet untouched.

Dubos shows that what we call "nature" today is a humanmade, or at least, a humanaffected environment. True wilderness, untouched by human hand, simply has not existed for millennia except in the most remote regions of the Antarctic and the highest mountain ranges. According to Dubos:

Humans stopped living a natural life more than ten thousand years
ago, not fifty or one hundred years ago. The days when humans
stopped living solely by hunting and gathering and began to construct houses or shelters, to cultivate wheat or rice, and even more
to build fires for warmth, they began to leave the natural life behind
. . . We still have the capacity to live naturally, but today we never
live in a natural state, because what we call the Temperate Zone is
incompatible with human physiology. If it were "natural" here, well,
the room where we're now sitting would be in the middle of dense
trees and probably marshes.

We do not live in an either/or environment, with nature on
one side of the fence and human civilization on the other. For
tens of thousands of years, human beings have been altering
nature. Think about it, and look at the environment around you
through Dubos's eyes:

In the temperate zone, a typical humanized landscape consists of
pastures and arable lands in the low altitudes and on gentle slopes.
Forests occupy almost exclusively the higher altitudes and other
areas unsuited to agriculture, industry, or human habitation. Most of
the bodies of water have been confined within well-tended banks,
controlled by dams, re-channeled, or disciplined in other ways. Despite all this human ordering, we forget that these typical sceneries
bear little resemblance to what they would be without human management. We have lived in intimate familiarity with them so long
that we contemplate them in a mood of casual acceptance and
reverie without giving thought to their origin and evolution. We
even forget that most villages and cities are on sites first occupied
by human settlements centuries or millennia ago and that roads,
highways, and railroad tracks commonly follow trails first opened by
hunters, pastoralists, and farmers ages ago.

In those ages long ago the temperate zones of the world were
covered with forests. We sense echoes of the danger and dread
that our forebears felt when we recall the dark, mysterious
forests of childhood tales. It was a dark and dangerous world for
the earliest hunters, the earliest herdsmen, the first farmers.
Even in Arthurian times, after the Roman Empire had ruled the
British Isles for centuries, Britain was covered with spirit-
haunted forests.

Over the millennia, over some 50,000 human generations, we have shaped this planet to our own purposes. The prize has been mastery of a world: The human race is supreme on Earth, unthreatened by any force save its own shortsightedness. The price we have paid has been overpopulation that threatens to burst society's fabric asunder and drown us in pollution, war, and death.

We think of this planet on which we live as the only place where human beings can exist. The rest of the universe is "outer space," quite separate from us.

Either/or.

Yet two human beings are living in "outer space," aboard the Soviet Union's Salyut 6 space station, as I write these words.

Ask a physicist whose specialization is the study of the upper atmosphere where "outer space" begins. He or she will give you what seems like a vague, almost evasive, answer. There is no boundary, no demarcation line, no precise altitude at which Earth's atmosphere ends and space begins. Instead, the physicist will give a range, a region, a set of numbers that says, in essence, "In this region Earth's atmosphere gradually thins out to the point where it becomes indistinguishable from 'outer space.'"

Scientific hairsplitting? No. Earth is a part of the Solar System. It is not fenced off from the rest of the universe.

Ask a solar astrophysicist and he might even claim that Earth orbits *inside* the Sun! That seems patently absurd. We can see the Sun; it has a sharp edge to it. Any astronomy textbook will tell us that the Sun averages some 93 million miles (about 150 million kilometers) away from us. Yet to a physicist who studies the Sun and knows that its outermost wisps of plasma stream past us at supersonic velocities, we can be considered to lie within the Sun's physical structure. It is very tenuous, this Solar Wind, a vacuum by human standards. But it exists, we are wrapped in it, and it affects the Earth in many subtle ways. The Northern and Southern Lights are visible manifestations, among others, of the fact that we are indeed touched by the Sun.

Our planet does not exist in isolation. The Earth is one planet out of many, in a Solar System that is rich in energy and natural resources. The sunlight that falls on Earth is only a minuscule fraction of the energy that the Sun pours into space. We know from the lunar samples returned by our Apollo astronauts that the Moon is rich in aluminum, titanium, oxygen, carbon, silicon.

For more than a century, astronomers have been studying the meteorites that flash through our atmosphere from deep space and crash into our planet. From their analyses of the contents of these chunks of stone and metal, they have concluded that hundreds of millions of tons of incredibly valuable metals and minerals are floating in our Solar System: iron, nickel, cobalt, platinum, gold, silver, potassium, carbon, magnesium, copper — hundreds of millions of tons of each of those elements. And much, much more.

Just as our ancestors faced the dark foreboding forests and tamed them to create the civilized landscapes of today, we face the cold and dangerous unknowns of space. That is the wilderness we have begun to explore, the new territory from which we can and must draw our sustenance.

To accomplish this, we must reject the either/or thinking that says we cannot fund a vigorous space program at the same time we fund enlarged programs elsewhere in the federal budget. To pit the space program against social welfare programs, for example, is the cruelest kind of political ploy, disastrous for the entire nation, rich and poor, black and white. It is not *either* space *or* welfare. Without tapping the riches of the Solar System, all our welfare programs, all our national economy, will inevitably collapse in ruins. It is not *either* science *or* humanism, for science itself is an art form, the most human thing that human beings do. Without science and its offspring technologies civilization itself will die, and human freedom and dignity will perish in the ashes. It is not *either* nature *or* civilization, for human civilization is as much a part of this planet's natural environment as any tree or fish. And soon, if we are wise enough and bold enough, human civilization will be a part of the natural environment of the entire Solar System.

For if we are to solve today's global problems, we must look not only beyond the Earth, but also toward the future. Problems that have taken generations to accumulate require long-range solutions. It is imperative that we stretch our minds into the future, to see far enough ahead so that we can make plans that will succeed. Otherwise we are constantly in the position of attempting short-term, stopgap solutions that seldom work for long, if at all. It is this "Band-Aid" approach that has let yesterday's problems fester into today's crises and tomorrow's disasters.

We must look to the future in our struggle against humankind's ancient enemies of poverty, ignorance, and death. To stand pat and try to muddle through today's problems is to invite disaster. To yearn for a return to a sweeter, bygone era is to embrace catastrophe.

To solve the problems of poverty, we must create new wealth.

To solve the problems of scarcity, we must create new wealth.

To solve the problems of overpopulation, we must create new wealth.

To protect human freedom, we must create new wealth.

The new wealth that we seek is waiting for us, untouched and eternal, a few hundred miles above our heads.

12

Toward the New World

> I like the dreams of the future better than
> the history of the past.
> —THOMAS JEFFERSON

We must create new wealth.

We live in a Solar System that is incredibly rich in energy
and raw materials. More wealth than any emperor could
dream of is available for every human being alive, in interplanetary space. Instead of thinking of our world as a finite pie
that must be sliced thinner and thinner as population swells,
we must go out into space and create a larger pie so that everyone can have bigger and bigger slices of wealth. That is the
politics of plenty.

Look at the parallel with Europe and its discovery of the
Americas. In the fifteenth century, Europe was a smallish back-
water in world affairs. China had absorbed the Mongol conquer-
ors of a few generations earlier and had again become the most
powerful and civilized nation on Earth. The power of Islam had
engulfed much of the Balkans and swallowed almost all of the
Iberian peninsula. Central Asia and India were centers of rich
and vigorous cultures. Backed against the cold, forbidding At-

lantic, the Europeans were dependent on the East for silk, spices, and knowledge.

But the Europeans were developing something that all the wisdom of the East never quite came up with: science. The scientific method of thought. The organized, rational investigation of nature that had begun with the ancient Greeks centuries before the birth of Christ. Unfortunately, the rise of Christianity spelled the end of rational inquiry into nature. The first stirrings of scientific thought died. And, within a few generations, so did the high civilization of the Ancient World.

An intellectual interregnum of almost fifteen hundred years stretches between Aristotle and Galileo. God alone knows where we would be if those two giants of scientific inquiry had followed each other by only a generation or two, instead of fifteen centuries.

In 1543, though, Copernicus published his revolutionary treatise suggesting that the Earth is not the center of the universe. He waited until he was literally on his deathbed to publish it; as a civil employee of the Church, he well knew the furor his ideas would cause. (Even as late as 1839, when the town of Torun finally put up a statue to its most famous son, no Catholic priest would officiate at the unveiling.)

Copernicus was 19 years old when Columbus sailed across the Atlantic in his three tiny ships. Seeking the wealth of the Indies and China, Columbus found instead a rather unpromising land populated by strange primitive tribes. In time, he was brought back to Spain in chains, largely for his failure to deliver the riches he had promised. (In modern parlance, his funds were cut and his program canceled.)

Columbus had been trained in the explorer's art by serving for years as a captain under Prince Henry the Navigator, who sent a long series of exploratory probes down the coast of Africa, seeking a way around the continent that would open a trade route to India. Modern historians estimate that Prince Henry spent about the same proportion of Portugal's wealth on these expeditions as NASA spent of the US gross national product during the Apollo years.

Portugal found India. Spain found the New World. Using the highest technology of the time — deep-water sailing ships and gunpowder — Europe broke free of its locked-in position, hemmed against the ocean by the powers of Asia and Islam. Europe turned the ocean into a highway, a path to riches, instead of a barrier.

What did the Europeans find in their New World? At first, gold and silver. The wealth of the Incas and Aztecs. Fantastic legends about cities of gold were matched with realities as the *conquistadores* used European technology to shatter the Bronze Age civilizations of Mexico and Peru. Treasure ships laden with gold and silver plied the Spanish Main. The hard, durable, obvious wealth from these shining metals changed the world's balance of economic power. Smaller European nations — especially England and the Netherlands — joined in the plunder. Gold was wealth itself. Gold was power.

As permanent colonies became established in the New World, a subtler form of wealth began to cross back to Europe: tobacco, maize, the potato. Few people today realize what a truly revolutionary food the Incan potato has been. It changed the nutritional habits of all Europe, and the economies of nations from Ireland to Russia.

To create those colonies in the hazardous New World, a new business invention was created: the limited liability corporation. Colonization in the Americas was an enormously risky undertaking — and an enormously expensive one. Even the very wealthiest of men could lose their entire fortunes and end their lives in debtors' prison if a trade ship foundered or a colony failed to return a profit. So stock companies were formed, and a legal fiction created where the corporation itself became a "person" in the eyes of the law. The corporation might go bankrupt, but the investors' private fortunes were not at risk if the corporation did fail. Only the corporation's own assets could be attached by its creditors. Of course, if the corporation became successful, as the Virginia, Massachusetts Bay, Hudson's Bay, and East India companies did, the investors reaped huge dividends.

Within a few centuries the New World gave to the Old its most precious gift, a new social invention: the concept of large-scale democracy based on the legally guaranteed rights of the individual citizen.

Thanks in no small part to the technological developments of gunpowder and rifling, American yeomen proved the equal of British armies, and the North American colonies won their independence from the British Crown. (Two centuries later, Vietnamese peasants showed that they too could use modern weaponry to hold off foreign armies.)

The idea that people could rule themselves, without hereditary kings, spread quickly across the Atlantic. Revolutions caught fire in France, and then all through Europe. The revolutions being fought in Asia and Africa and Latin America today are stirred by the same impulses that moved our forebears to take up arms. Unfortunately, today we cast ourselves in the role of Redcoats and push revolutionary movements that could be Jeffersonian into the arms of the Marxists.

*

What can we gain from the New World that begins a couple of hundred miles above our heads? There is gold out there. More than any Incan Emperor ever knew. And silver, platinum, diamonds, iron, aluminum, copper — any metal or mineral that exists on Earth is waiting for us in space, by the thousands of millions of tons. No one owns it. Yet. Of more immediate concern, there is energy in space. Enormous energy from unfiltered sunlight. The Sun radiates an incomprehensible flood of energy into space: the equivalent of ten billion megatons of H-bomb explosions *every second*. The Earth intercepts only a tiny fraction of this energy, less than two ten-millionths of a percent.

Pioneer thinkers like Konstantin E. Tsiolkovsky, the founder of Russian astronautics, and Freeman Dyson, of the Institute for Advanced Study at Princeton, have envisioned a far-distant future when humankind has built an energy-trapping shell around the Sun so that *all* of our star's titanic energy flow can be used by the human race.

For the purposes of our generation, though, a few dozen Solar Power Satellites in geostationary orbit above the Equator, may be sufficient. Each such SPS could beam five gigawatts (billions of watts) to Earth; that would make a substantial contribution toward solving our energy — and, ultimately, our economic — problems.

Energy is there in space, in superabundance. We have merely to decide to reach out and tap it.

Natural resources are there, also. Also in superabundance. Thanks to the Apollo explorations of the Moon, we know that lunar rocks and soil contain many valuable elements such as aluminum, magnesium, titanium, silicon, oxygen. Lunar materials could support human miners, explorers, scientists, and construction engineers on the Moon, or elsewhere in space.

Looking further afield, there are the asteroids — chunks of rock and metal — which sail through the Solar System, mostly in the region between Mars and Jupiter that has been dubbed the Asteroid Belt. An astronomer once called the asteroids "mountains floating free in space." To a miner or an industrialist they look more like a bonanza. One-kilometer-long asteroids are not unusual. With Earth-based telescopes, more than 2000 asteroids are big enough to be observed regularly, and there are probably 100,000 or more that our biggest telescopes could detect out at the distance of the Asteroid Belt. MIT astronomer Tom McCord estimates that there are hundreds of millions *of billions* of tons of nickel-iron asteroids in the Belt. The economic potential of these resources is incalculably large. And mixed in with these are many other valuable metals and minerals, from aluminum to zinc, from water to gold.

The idea of getting our world's steel supply from an asteroid has a strange, mystical resonance to it. For the very first artifacts that human beings made of iron, the archeologists tell us, were probably cold-worked from meteoric iron. Stone Age peoples did not dig iron out of the ground, but they occasionally found oddly pitted "rocks" and used the metal to make implements and ornaments. We started on the road to metals technology with meteoric iron. Now we can go to the source of those meteorites, to the Asteroid Belt, and supply the whole world's needs for iron, steel, and thousands of other metals and minerals.

This "mother lode" of riches is a long way from Earth: hundreds of millions of miles. But in space, distance is not so important as the amount of energy you must expend to get where you want to go. Like the ships that sailed the Spanish Main, spacecraft do not burn fuel for most of their flight; they coast after achieving sufficient speed to reach their destinations.

Space is not a barrier. It is a highway. The biggest and most difficult step in space flight is getting off the Earth's surface. A spacecraft must expend enough energy to achieve a velocity of some 18,000 miles per hour to reach a low earth orbit (LEO). But once in LEO, the hardest work has been done. To go from an orbit of about 200 miles' altitude to a soft landing on the Moon, a quarter-million miles away, takes only about 75 percent of the energy it cost to reach LEO.

To travel the hundreds of millions of miles to the Asteroid Belt will require less energy, per pound of payload, than it did to boost the payload up from the Earth's surface. But it will take time: months or even years. Like the New Bedford whalers who set out to sea on voyages of three or four years, tomorrow's prospectors and mining engineers will sail off into space on multi-year missions, with multi-year employment contracts in their pockets.

As we set off into the depths of space, as we begin to send our workers (men, women, machines) outward through the Solar System, we seem to be obeying a basic human urge to explore, to extend our habitat, to reach out beyond the unknown.

The human species began in warm savannahs in Africa and

Asia some two or more million years ago. Through physical adaptation we developed grasping hands, stereoscopic vision, speech, and the intelligence that has made us the masters of this planet. Since the Ice Age of the past million years we have explored and expanded until now we live on every land surface on Earth, from pole to pole. Even in the most desolate dry valleys of Antarctica, where not even viruses are found, human explorers live.

Most of this global expansion has taken place in the 500,000 years since we tamed fire, the Promethean gift. We have developed our technology from sticks to pebbles to metals to coal to oil to uranium and silicon electronic chips. During these 500 millennia our physical form has changed relatively little. Even today, down inside the chemistry of our genes, we are genetically so similar to chimpanzees that biologists are astounded by the physical and cultural differences between us and the chimps.

Our physical form has changed little. Our social organization has changed vastly. This is the way that we change today: not physically but socially, culturally, technologically. Physical changes are slow; they require millennia, eons. But we change, adjust, alter our society in a generation or even less. This is the secret of humankind's success as a species on this planet. We no longer change physically to adapt to a new environment; we use technology to bring our environment along with us, wherever we go.

Just like any other biological organism, we are driven to expand our ecological niche as far as we can. Only in this way, our genes tell us, can we assure our survival. So we move all across this planet's land surfaces, into deserts and jungles, out onto islands and ice floes, down to the bottom of the sea — and out into space.

The physical adaptations of our bodies to harsh environments are minuscule compared to the social and technological adaptations we have made. The elaborate social customs and clothing styles of the Eskimos and Bedouin Arabs show how far we can adapt to harsh environments. The Eskimo's parka and the Bedouin's robes are both technological adaptations to the environment. The human body has not been altered to survive Arctic cold or desert heat: That body still dies if its internal temperature strays a few degrees from 98.6° F.

Through our technology we create a microenvironment that carries our ecological niche along with us, wherever we travel. In the blazing aridity of the desert, in the numbing chill of the Arctic, at the bottom of the ocean, in the depths of space, we extend our ecological niche into new domains by taking our environment along with us. Thus we can expand into new domains without changing physical form.

We move outward into space because it is biologically necessary for us to do so. Like children driven by forces beyond understanding, we head into space citing all the good and practical and necessary reasons for going. But we go because we are driven. The rocket pioneer Krafft Ehricke has likened our situation on Earth today to the situation in a mother's womb after nine months of gestation. The baby has been living a sweet life, without exertion, nurtured and fed by the environment in which it has been enclosed. But the baby gets too big for that environment, and the baby's own waste products are polluting that environment to the point where it becomes unlivable. Time to come out into the real world.

Our biological heritage, our historical legacy, our real and pressing economic and social needs are all pointing toward the same conclusion. Time to come out into the real world.

Time to expand into the New World of space.

III

THE PROGRAM

Who dares nothing, need hope for nothing.
—FRIEDRICH SCHILLER

13

Future Three

The year is A.D. 2000.

Civil War has begun in the United States. On one side, the ghetto dwellers of the big American cities: mostly black, brown, yellow. Against them: the affluent middle class of whites and blue-collar workers who live in the suburbs and the farmers. In the middle: the federal government (mostly white) and the US Army (mostly black).

Mediation has failed. The widening gap between the rich and poor finally snapped when Congress pushed through the Solar Power Law over the President's veto. The law called for subsidizing the development and deployment of solar energy systems for private homes by putting an additional federal tax on imported oil and domestic coal purchases. Added to the recently passed Clean Air Act of 1999, which makes coal burning prohibitively expensive and sharply limits the operation of the few remaining nuclear power plants, the citizens of the city ghettos finally erupted into rebellion.

"Why should we pay for *their* comforts?" demanded one of the ghetto leaders. "Cities can't run on solar power and now Whitey won't let us burn oil or coal. Hell no, we won't pay it!"

Demonstrations in every major city quickly turned into riots, and when the Army was called in to restore order, most of the black soldiers refused to fire on their brothers. Instead, they joined the rebellion.

The Navy and Air Force appear to be on the side of the whites, although it is too soon to tell. Suburban and rural whites are organizing their own militias and vigilante squads. In many cases these are better armed than the Regular Army units, although of course the Army has tanks and helicopters. There have been reports of massacres of scattered communities of Indians, Mexicans, even Chinese, throughout the rural West.

President Martinez has warned the Soviet Union that any attempt to intervene in the internal affairs of the US will be met by a full-scale nuclear strike. The Strategic Air Command and the Navy's Trident missile submarine force have both sworn allegiance to the President. The President also announced postponement of the November elections until such time as the existing emergency subsides.

The Russians, meanwhile, are rumored to be deploying laser-armed satellites that can shoot down missiles long before they reach their targets. Several of our early-warning satellites have suddenly become incapacitated. What effect this will have on the situation is difficult, if not impossible, to assess.

14

What Elephant?

> The impossible we do immediately; the miraculous takes a little time.
> —INFORMAL MOTTO OF THE US
> NAVY CONSTRUCTION BATTALIONS
> (SEABEES) OF WORLD WAR II

In a broadway show titled *Jumbo,* more than half a century ago, Jimmy Durante walks on stage leading a full-grown elephant on a leash.

An actor asks him, "Where'd you get the elephant?"

"What elephant?" snaps Durante.

Every day, so many aspects of our lives are affected by the satellites orbiting overhead that we tend to forget they are up there, patiently working for us. We take for granted the savings in money and lives that these space artifacts have given us. We assume that the additional comforts and protections they provide us are part of our ordinary, natural environment.

What elephant?

Take a closer look at the television news broadcasts. Right there, every night and day, satellite photographs show you the sweep of weather patterns across the entire North American continent and far out at sea. Live and taped news reports are relayed daily from all over the world to your local transmitters by communications satellites.

Ordinary, everyday things — which did not exist until 1960 or later.

Just outside Lawrence, Kansas, where Quantrell's Raiders made their infamous raid during the Civil War, there stands a sleekly modern building across a busy highway from the University of Kansas. It is the Kansas Space Technology Center. Without much fuss, students and professional staff there pass on the benefits of the space program to the farmers, ranchers, city planners, real estate developers, and taxpayers of the state — every day.

A century ago cattle ranchers hired John Wayne–type cowboys to guard their herds, patrol the open rangelands, check where woods and scrub growth were encroaching on the pasture grass, keep watch on the amount and quality of water available for the herds. Today, in the single click of a camera shutter aboard a satellite orbiting a few hundred miles above, all that information is recorded for the entire state of Kansas, as well as all of Kansas's neighboring states. Farmers need early warning of crop diseases, insect infestations, information about water table levels. Satellite data, processed at the Space Technology Center and distributed throughout the state, provides such information.

While almost every other aspect of modern society is driving the price of food up, space technology is providing information that helps prevent the price of beef and grains from rising even higher. No one has yet done a detailed calculation of cost and benefit, but the chances are that the $50 million or so it costs to launch a Landsat–type of observation satellite is repaid, with dividends, by the cost savings in food products that the satellite provides.

The same Landsat spacecraft help city planners to see how land area is being used, where new housing developments are spreading, where the factories are, where the best areas for new developments may be. Throughout the entire state, in one click of a camera.

City councils receive satellite photographs that pinpoint sources of air and water pollution. Fishing fleets are guided to schools of fish far out at sea by satellite information. Aircraft navigate around the globe, ships are never out of touch with navigational information or communications links. All because of those satellites we take for granted.

During the severe winter of 1977–78 the citrus fruit growers of Florida were saved some $45 million in crop damage over one night by satellite warnings of a coming frost. The growers knew that a frost was going to strike, but *when?* Pinpoint information from a weather satellite told them exactly when they needed to light their smudgepots. The savings in fruit and in fuel (since they did not have to start the smudgepots sooner than necessary) paid for the cost of the satellite in a single night.

The National Oceanographic and Atmospheric Administration estimated that the nation saved $172 million *over and above the costs of launching weather satellites* during the year 1978 alone.

Communications companies routinely pay NASA to orbit relay satellites for them. Even at $50 million per launch, satellite relays are cheaper by far than stringing networks of relay towers across a continent or laying cable across an ocean.

Intelsat, the international communications organization to which 105 nations now belong, handles about 60 percent of all transoceanic telephone calls through its satellites. Almost all intercontinental television broadcasts are relayed by satellite. Almost 150 countries and territories use satellites for international communications, and large nations like India, the US, and the USSR use satellite relays even for phone services inside their own borders.

In the United States, the Comsat Corporation was formed during the 1960s as a mixed public/private company jointly controlled by the government and private corporate members (and selling stock on the open market to any investor who wants to buy it). Comsat has entered into a consortium with IBM and Aetna Life & Casualty Company to create Satellite Business Systems (SBS).

Two SBS satellites will relay business documents, correspondence, and computer data directly from office to office via rooftop antennas at speeds of up to 20 times the speed possible with land-based communications links. The first SBS satellite was launched in November 1980.

Over the coming decade, at least 25 percent of all long-distance voice communications and half the computer data and television traffic in the world will be relayed by satellites. Market surveys show that in sub-Saharan Africa alone, between 8000 and 20,000 new ground stations will be needed to handle the satellite communications load of the foreseeable future.

By the end of this century, our most vital information links for communications, business, defense, and personal uses will depend on relay satellites. From Wall Street to Main Street, we all use satellites every day, without thinking about it.

When we make a phone call or watch television or plan a picnic or vacation or hear a storm warning, we are being served by satellites. The prices we pay for food would be even higher than they are if it weren't for satellites. And people who live in areas threatened by hurricanes owe their lives to storm warnings from satellites. Certainly such warnings have saved property owners billions of dollars of damages.

In their own way, satellites have even helped to make this nuclear-armed world a bit safer. And they may be a vital key to reducing the titanic tax burden of modern armaments.

Before the advent of satellites, every effort to negotiate an arms limitation treaty between the US and USSR ran aground on the same issue: on-site inspection. American negotiators insisted on the right to inspect Soviet military installations and count the number of missiles emplaced. Without such a safeguard, no reasonable treaty could be agreed to. The Soviets flatly refused to allow "spies" on their territory. But even as the Russians were being intransigent, they were themselves providing the means for breaking the deadlock — without knowing it.

They launched Sputnik.

When they lofted the first Sputnik into orbit, October 4, 1957, the Russians never asked anyone for permission to fly over their national territory with a satellite. In 1957 national sovereignty was legally defined, by international agreement, as extending upward from a nation's borders to infinity. The legal theory was that each nation had sovereign rights to whatever portion of outer space that happened to be above the nation's terrestrial territory. The theory was blithely accepted by everyone as long as no one had a spacecraft in orbit. The real purpose behind the international understanding concerned aircraft and airspace: No nation wanted foreign aircraft buzzing over its landscape without prior permission.

Then Sputnik went into its highly inclined orbit and crossed the borders of almost every nation on Earth. The international lawyers were even more shocked than the rest of the world at the Russians' feat. The Russians had not asked for permission to overfly other nations! The Soviets simply stood pat with their *fait accompli.* After all, what could anyone do about it?

But the Russian attitude abruptly changed a few years later, when the US started launching military observation satellites. "Spies in the sky!" thundered the Soviets. The Americans smiled politely and pointed out that Sputnik had established a legal precedent: You could orbit over any nation's territory without prior consent. There was no legal way that the Soviets could prevent the US, or any other nation, from sending satellites over Russian soil. Their only recourse was to launch their own reconnaissance satellites, which they soon did.

Law always lags behind technological developments, but to their credit the international lawyers quickly hammered out a new understanding. National sovereignty still holds for the *air*-space above a nation's territory. But outer space, above the altitude where an airplane can operate, is an international regime, legally. Anyone who can get there can operate in space. We will see in a later chapter how the development of legal agreements and international treaties affects the space efforts of the United States and other nations. Some nations are beginning to claim sovereignty over choice orbital sites. The United Nations' so-called Moon Treaty threatens the very existence of the American space program.

But in the 1960s, with both sides snooping on each other from orbit, the way was suddenly clear to inspect missile build-ups without setting foot on foreign soil. As the Cuban Missile Crisis of 1962 proved, high-altitude reconnaissance can spot missile emplacements rather easily.

The first Strategic Arms Limitation Treaty (SALT I) was negotiated with the knowledge that both the US and USSR would use satellite photography to verify the numbers of missile silos and missile-armed submarines being deployed. That first small step back from the nuclear arms race could never have been accomplished without satellites. Today satellites patrol the skies, watching with the stolid patience of machines against missile build-ups, launches, and nuclear bomb tests on Earth or in outer space. In the bars around the Pentagon you can hear bantering references to satellite photographs so detailed "you can read the license plate numbers on the cars in the Kremlin parking lot."

American intelligence estimates of the size of the Russian wheat crop depend almost entirely on satellite data. These estimates play a key role not only in our political relations with the Soviet Union, but in business decisions on how much of *our* wheat crop the Russians are likely to need in the coming year.

Gilbert W. Keyes, deputy manager of Strategy Planning and Market Development for Boeing Aerospace Company, points out:

> Anyone who has some insight on the operations of the giant "crop futures" markets in the American midwest, and understands how highly prized crop forecasting data has become for these global traders, will recognize the business potential in this area. Even more

important is the fact that these [satellite] capabilities will enable us to avoid buying panics when global food supplies run low, as in 1973/74, and help us to penetrate the statistical fog around Soviet crop production that enabled them to make such favorable grain purchases in 1972.

As the world intensifies its efforts to grow food for our burgeoning global population, satellite data on weather, water tables, rainfall expectations, ground use, crop growth, and health become ever more crucially important. Yet this very information may not be made available because of political and legal squabbles. Many national governments do not like the idea of a satellite (especially an American satellite) photographing their territory and then making the data available to the international community. This is not merely an East-West problem. Most of the poor nations of the Third World balk at having satellite data about their territory given out without their permission. Publicly, they express fears that data revealing new areas of natural resources will fall into greedy hands. Privately, they resent having the rest of the world see exactly how poor they are. Thus the development of a global information network disseminating data about food crops, natural resources, pollution, weather, land use, and so forth is stymied by a form of nationalistic space-age paranoia.

Satellite data can make a vital difference in the amount of food available to feed the world. Satellites can help to locate new sources of natural resources and pinpoint sources of pollution. But these opportunities are now hog-tied by legal red tape. The result: The rich get richer, because the industrialized nations utilize satellite capabilities to the utmost. The poor, who could benefit most from such help, get poorer because their governments do not trust the rich and fear being embarrassed in the international community.

Communications satellites are a much different, much brighter, story. Since the first Telstar was orbited in 1962, comsats have grown into a billion-dollar-per-year industry, with corporations such as Western Union, AT&T, RCA, and many others actively involved. Even among the poorer nations of the world, communications satellites are having an impact. Thanks to Arthur C. Clarke, who invented the idea for comsats in the first place, a cooperative program between the US and India provided the first major test for *direct broadcast satellites*. These are satellites that beam their signals directly to individual receivers in people's homes or in public buildings.

Antennas for direct broadcast reception are relatively cheap: Neiman-Marcus advertises them for a few thousand dollars. Electronic hobbyists can build them for much less. With such an antenna on your roof, you can pick up the signals beamed by the satellites directly, without having to tune to a television station.

Clarke, who lives in Sri Lanka (formerly Ceylon), conceived the idea of placing direct broadcast receivers in every village in India. The satellites could then beam educational programs throughout the subcontinent. In a stroke, he reasoned, a millennium's worth of ignorance could be lifted from the backs of hundreds of millions of India's poor villagers. The United States provided the satellite. Cheap antennas sprouted throughout the villages of India. The technology worked fine. But somehow the Indian government seemed to beam more political propaganda and entertainment movies to the villagers than educational material.

The politics, too, lag behind the technology.

Lest you think that this is a particular problem of India, or of developing nations in general, let me tell you about my first visit to Ted Turner's "superstation" in Atlanta. The flamboyant Mr. Turner is a high-powered businessman and America's Cup yachtsman. He has created, through modern electronics and satellite relay, a mini-network that beams his Atlanta-based TV transmissions all across the North American continent, from Alaska to Puerto Rico. It's called the TBS Network, and the television industry regards it as the shape of things to come. But what was this sophisticated equipment beaming out to its satellite, for relay to millions of homes all across the continent? Reruns of *I Love Lucy,* followed by *Leave It to Beaver.* Space technology has already done a lot for us and repaid every cent we have spent on the space program, with interest. Yet it seems fair to ask if it's done enough. After all, "From those to whom much is given, much is expected." Even with all the profits and benefits already flowing to us from space, we live in a world that is hungry and tense, in an America that is torn by economic and social problems. Can space technology do still more to help solve problems of energy, economic growth, employment, social unrest? Yes, unquestionably. The next few chapters will show some of the things that can be done *now.* Not in the far-flung future. Not in our children's time. Now. Space projects can produce jobs for us, today, and profits for companies and individuals wise enough to invest in space.

Space technology can help solve those economic and social problems that seem to be so far beyond the grasp of the politicians and lawyers. Space projects can return enormous benefits to each of us: energy, raw materials, new industries, new jobs, better and safer lives.

What elephant? As the great Durante would say, "We've got a million of 'em!"

15

The Once and Future Technology

It's no use pretending that we can live in the
twentieth and twenty-first centuries without a vigorously advancing technology.
—ARTHUR KANTROWITZ

We have not always used technology to our best advantage.
Short-term profits have often (some say always) received more
attention than long-term benefits. We tend to think of immediate comforts rather than eventual needs. Sheer laziness, physical and mental, often gets the upper hand over careful planning.

The fact that we can utilize the knowledge gained from space
technology to create a second-generation technology that will
provide for our physical needs without destroying the environment does not necessarily mean that we will do so. It's so much
easier to go on with business as usual, to drift and delay all
changes as long as possible. Today's problems are real; tomorrow's benefits are theory; and besides it will cost too much to
implement.

The fact that we can reach out into the solar system and bring
back incalculable treasures of energy and natural resources to

the needy billions of Earth does not guarantee that we will do so. It takes an enormous amount of capital to venture into space. It takes vision and courage. The risks are high, the unknowns are many. Isn't it easier to tighten our belts and try to live through these cruel times of shortages, these lean years?

Let me tell you a story, a sort of science fiction story, about how we use technology to overcome adversity.

Once there was a technology called MHD. The initials stand for *magnetohydrodynamics:* the science that studies how ionized gases (plasmas) interact with magnetic fields.

MHD technology can be used to build a new kind of electric power generator that is very efficient. For every pound of coal or oil or natural gas burned, an MHD generator can yield 50 percent more electricity than conventional power generators. In addition to being efficient, MHD generators can be very clean. They can burn the dirtiest high-sulfur coal, for example, and yet emit little more than carbon dioxide and water vapor from their smokestacks.

The first MHD generator was a modest affair that produced 10 kilowatts for a few seconds at a time. It was a laboratory model, intended to show that MHD power generation was possible, nothing more. That was in 1959.

By the mid-1960s, American scientists and engineers had built MHD generators that produced more than 30 megawatts and smaller test generators that ran for hundreds of hours nonstop. We (for I was involved in this work, peripherally) were ready to build a full-scale demonstration power plant, a pilot plant, that would produce 50 megawatts from its MHD generator and would run for 10,000 hours continuously.

That took lots of money: at least $50 million, which was much more money in 1965 than it is today.

The pilot plant was a cantankerous beast, as pilot plants often are, and it took several years to get its MHD generator to operate reliably. But thanks to the foresight and daring of the corporation that owned the laboratory where the work was done, and the electric utilities that shared the burden of cost for the work, and the US government, the scientists and engineers ironed out the bugs in the pilot plant and got it to operate at about 60 percent efficiency — half again as efficient as the best power generators in use at that time.

With uncommon boldness, American industry pushed ahead and began to build MHD power stations across the land, cheered by the fact that the power stations they were building would not pollute the environment and would provide electricity at cheaper rates than ordinary power generators could.

By a stroke of fate, the first commercial MHD power stations went "on line" late in 1973, just as the Middle East exploded with the Yom Kippur War that pitted Israel against Egypt and Syria.

When Saudi Arabia and the other OPEC nations slapped on their famous oil embargo, Europe and Japan were nearly paralyzed. Fortunately, the US had enough domestic oil production to get through the crisis — because the advent of MHD had made it possible for us to use coal for most of our electric power generation, thereby stretching our oil production for the place where it was needed most: transportation and heating. America could laugh at the Arab Oil Embargo, and did.

The Arab embargo quickly collapsed, and by the latter 1970s the US was selling MHD technology all around the world. The huge favorable balance of payments that resulted from this high-technology export made it possible to cut taxes drastically. Unemployment dwindled in the US, the standard of living rose, and even the angriest minorities found that jobs and affluence did more for them than demonstrations and lawsuits.

Lovely story, isn't it?

It's a science fiction story, a tale set in an "alternate universe" that is similar to our own, but clearly not the world we live in.

Up until 1965 that story was entirely true. MHD power generation is a reality. By the mid-1960s we had indeed produced MHD generators that put out more than 30 megawatts and we were ready to build that pilot plant.

But when the laboratory's parent corporation and the electric utilities we were working with realized that it would cost $50 million for the pilot plant, they turned their backs on MHD. None of the "energy corporations" (oil companies) had the slightest interest in backing a new technology that would burn their fuel more efficiently. The federal government was indifferent almost to the point of hostility.

The pilot plant got built, nonetheless. It exists today and it works just fine. It stands on the outskirts of Moscow.

The Russians, watching the laboratory work on MHD being done in America, started their own MHD program. Since Lenin's time the USSR has emphasized electrical power generation as a key to industrialization and progress. In this decade of the 1980s MHD will go "on line" in the Soviet Union, as it could have gone "on line" in the US in the 1970s.

MHD is a technology we threw away. Or, more accurately, we put it on a shelf. A library shelf. In the United States, MHD has produced more paper than kilowatts.

MHD is a space-age technology. It depends on combustion chambers that are very much like rocket engines and materials that can withstand searing temperatures of 5000° F or more. At a deeper level, the technology depends on understanding the flow of supersonic gases and the physics of high-temperature ionized gases, understandings that "spun off" from the earliest work in the re-entry problems for ballistic missiles and spacecraft.

The basic principles of MHD power generation go back to Michael Faraday, the son of a blacksmith who became one of the greatest experimental scientists of all time. In 1821 Faraday discovered that if you move something that can conduct electricity — such as a copper wire — through a magnetic field, you generate an electric current.

Sixty years later Thomas Edison used this knowledge to start the electrical power industry. Edison's power generators were built along the same principles as Faraday's lab-bench "dynamo." A bundle of copper wire, called an *armature,* is spun as rapidly as feasible within the magnetic field created by powerful magnets. To spin the armature, Edison used a steam turbine. To spin the turbine, he used a coal-fired boiler and made steam.

Edison's power plants were about 40 percent efficient. Today, a century later, modern power plants still boil water to make steam to turn turbines to spin armatures. And they are still about 40 percent efficient. Except for nuclear power plants, which are deliberately run at lower efficiencies for safety reasons. In a nuclear plant, the heat of the nuclear fissions takes the place of the heat from burning a fossil fuel. But it all still goes into boiling water to turn the turbines.

When Faraday discovered the principle of electric power generation, he realized that *any* material that conducted electricity could be used, in theory. A bundle of copper wire was the most practical because copper is an excellent conductor of electricity.

Faraday even tried to measure the electric current generated by the flow of the River Thames as it moved through the Earth's geomagnetic field. The current was too weak for him to measure, however.

An ionized gas, which physicists call a plasma, can conduct electricity. This is the heart of the MHD idea. Although most plasmas are much poorer conductors of electricity than copper, you can roar a plasma through a magnetic field at supersonic velocity, thereby making up in velocity what you lack in conductivity.

An MHD generator can do away with all the mechanical moving parts that Edison and his heirs needed. No steam boiler. No turbine. No drive shaft. No armature. Nothing but a burner with a rocket-like nozzle on it to blow hot combustion plasma down a pipe. Wrapped around the pipe is a powerful magnet. Inside the pipe are electrodes that tap off the current. That's all there is to an MHD generator.

The basics are simple. The technology is very difficult, especially the materials technology which must face up to very high gas temperatures, very powerful electrical voltages, and very corrosive pollutants in the plasma.

A typical MHD generator runs at 60 percent efficiency, half again as good as the generators that produce our electricity today. The engineers claim that the MHD system can be greatly refined and made even more efficient than these early machines.

The pollution products inside the MHD generator are so thick that they can be tapped out, using well-known techniques of gas separation, and processed into chemical fertilizers rich in sulfur and nitrogen compounds. In a sense, the pollution inside the MHD generator is so bad that it's good. It becomes profitable to extract the pollutants long before they reach the smokestack.

MHD generators running on coal must be fitted with stack scrubbers to get rid of the particulates (soot) that inevitably accompany coal burning. But here again, the very high temperature at which the MHD system runs may create less soot than the lower-temperature combustion of standard turbine-generator systems.

The emissions from an MHD power station's stack should be very clean carbon dioxide and water vapor, even if the plant is burning high-sulfur coal. Of course, the MHD system can also burn oil or natural gas. To date, existing commercial nuclear reactors are not designed to produce the high-temperature heat that an MHD generator requires, so the mating of MHD with nuclear reactors seems rather distant.

If, as in our fiction scenario earlier, the US decides to turn a major portion of its electrical power generation capacity over to coal, MHD may be the only way to generate the electricity we need without fouling our air beyond redemption. Turning the coin on its other side, you can see that MHD offers us the opportunity to use the coal we have in such abundance with some assurance of environmental protection.

This does not mean that MHD will work as advertised, and that all we need to do is start building the power stations. There are still unknowns. But if we had built that pilot plant in 1965 and followed up on the work that had already been done, we would know for certain today whether or not MHD technology could be counted on to help.

We don't know, and we should find out.

The MHD program in the USSR is proceeding at a pace much slower than American enthusiasts would have predicted. Is this because of the inherent difficulties with the technology, or political decisions within the USSR, or simply that the Soviets cannot move as fast as we can — once we make up our minds to move?

It's instructive to look at the spread of microelectronics technology, such as pocket calculators and microcomputers, when comparing US and Soviet progress rates. In the socialist nations, microchip technology is spreading slowly, even though the technology is sorely needed throughout all their industries. In the capitalist nations of the West, spurred by unabashed greed that produces waste, white-hot competition, and quick profits, microchip technology has swamped everything from the automobile industry to the games industry. Home computers, computer games, pocket calculators, *wristwatch* calculators that also play music, microprocessors in sewing machines, typewriters, everywhere — microelectronic gadgets are flooding Western society.

When it was roaring along at its best, back in the early 1960s, the American MHD program was privately funded. Now the US has an official, government-funded MHD program, and many of the old-timers who remember the sixties say that the US program now looks much like the Russian one: a bureaucratic paper mill. More money is being spent on MHD in the United States today than ever before. But the program is inching along, spread thinly among dozens of universities and private laboratories.

MHD technology can be crucial to our decisions on how, or if, to use coal as our main energy source for the near and middle-term future. With or without MHD, coal can be only a stopgap in our long-range energy plans, because of coal's severe environmental effects, which range from carcinogens to lung infections to acid rain that destroys life in lakes and streams and ruins timberlands and farms. The current level of acid rainfall over New England alone could cut that region's productivity in agriculture and forestry by ten percent over the next decade — a loss equal to the energy output of fifteen one-megawatt power stations.

Even with MHD to alleviate the immediate pollution problems associated with coal burning, fossil fuels pour carbon dioxide into the air, and this builds up the greenhouse effect that heats the atmosphere. The long-term climatic impact of coal-burning is unknown, but it is bound to be severe.

The maddening thing is that we could have given MHD a fair trial by now and seen for ourselves where it could fit into our energy plans. It may be vitally necessary to have MHD if we must turn to coal.

If that happens, the chances are that we will have to buy our MHD technology from the Russians — if they are willing to sell it back to us.

16

Nuclear Power: The Bad News

> If you once forfeit the confidence of your
> fellow citizens, you can never regain their
> respect and esteem.
> —ABRAHAM LINCOLN

Nuclear power is a fact. Like it or not, nuclear power stations exist not only in the United States but in many other nations as well. The question is: What share — if any — will nuclear power have in the world's future energy scene? Allied to that question is another: Assuming we will continue to use nuclear power, how can space technology help to make nuclear power plants safer, more reliable, more efficient?

The Luddite answer to the problems concerning nuclear power has the simplicity of madness: "No more nukes!"

The Prometheans, shaken by the failure of nuclear power to fulfill all the dreams of its founders, reply that nuclear power cannot be dispensed with; we need it, now more than ever, and we know that it works and is safer — repeat, safer — than many other forms of energy.

But the Luddites are on the attack, and they smell blood. Critics such as Amory Lovins insist that nuclear power stations are too risky, too expensive, too little, and too late to help us solve our energy problems. They fear the dangers of catastrophic radiation leakage and point out (quite accurately) that nuclear power plays only a minor role in the total US energy picture.

The attack is extremely emotional. Not only are scientists like Barry Commoner profoundly antinuclear; the "No Nuke!" chorus includes men and women from all walks of life, of all ages and income levels. Jane Fonda's antinuclear speeches get more attention from the media than the pronuclear utterances of Nobel Laureates such as Hans Bethe or Glenn Seaborg.

Comedian Dick Gregory puts aside his jokes when he speaks about nuclear power. Once he vowed to eat no more solid foods until *every* nuclear power plant is closed down. In a speech in Connecticut in 1980, he solemnly stated that all the nuclear plants in the US are built within two miles of an earthquake fault, and that the accident at Three Mile Island was no accident at all. (In the same speech, he also said that women are being sterilized by pantyhose and that the government gave syphilis to black men by giving them vitamins.)

The Luddites insist that the expense and hazards of nuclear power will prevent nukes from playing a major role in the US energy picture. This seems to be a self-fulfilling prophecy. Thanks in large part to the Luddite offensive against nuclear power, nuclear reactor orders in the US dropped to zero in 1978 — well before the Three Mile Island accident of March 1979. In the five years between 1969 and 1974, 87 reactors had been ordered; between 1974 and 1978, the orders went down to six. Not one has been ordered since 1978.

Yet in other nations, such as France, Japan, the Soviet Union, nuclear power is being pushed hard. Take France, for example. Seventeen percent of that nation's electricity now comes from nuclear power plants, by 1985 it will be 55 percent, and by the year 2000, 80 to 90 percent of France's electricity will be generated by nuclear power plants, if existing plans are carried through. Despite strong protests, the French government is implementing a complete nuclear program that has been compared to the American Apollo space effort of the 1960s.

France must import all its oil and most of its coal, so nuclear energy is an attractive alternate. French officials estimate that nuclear-generated electricity costs about 60 percent less than oil-generated electricity and a third less than that generated by coal.

The power of the atomic nucleus was harnessed under the terrible lash of wartime urgency. Many of the physicists who worked on the Manhattan Project were Europeans driven from their homes by Hitler, bitterly aware that millions of innocent men, women, and children were being systematically slaughtered by the Nazis. They knew also that German scientists were working on the problem of splitting the atomic nucleus and feared that the Nazis might develop nuclear weapons before the Allies could.

Robert Oppenheimer, the scientific director of the program, said years later of the Manhattan Project:

> When you see something that is technically sweet, you go ahead and do it and you argue about what to do about it only after you have had your technical success. That is the way it was with the atomic bomb.

So the men and women of the Manhattan Project had their "technical success," and Hiroshima and Nagasaki were the result. The scientists were appalled. They had appealed to President Truman not to use the bombs on civilian targets, but rather to demonstrate the awesome power of a nuclear weapon in some way that would convince the Japanese to surrender.

Truman decided the only demonstration that would work would be against a real target. More than 100,000 casualties were inflicted at Hiroshima. Although much worse devastation and higher casualties had already been caused by the fire raids on Tokyo and Osaka by B-29's carrying incendiary bombs, those raids were the work of hundreds of aircraft dropping thousands of tons of bombs. Hiroshima was wiped out by one airplane, in an eyeblink.

The Japanese surrendered, saving us and themselves the prospects of a bloody invasion of their home islands. The most sanguine American military estimates of the cost of such an invasion were that it would inflict more than a million casualties on American troops, and perhaps five times that number or more on the Japanese population.

The nuclear bombs were less destructive than the fire raids and saved both sides the incredible havoc of invasion. Yet the bombs stunned the world and deeply wounded the consciences of many of the scientists who had created them. Oppenheimer said in 1948, "In some sort of crude sense, which no vulgarity, no humor, no overstatement can quite extinguish, the physicists have known sin; and this is a knowledge which they cannot lose."

Not all the physicists felt the responsibility for Hiroshima and the resulting strategic "balance of terror" between the US and USSR as deeply as Oppenheimer. Certainly none of them was so eloquent about it.

But it was a profound sense of moral obligation that drove many American scientists to try to beat their sword into a plow-share, to turn the power of the atomic nucleus into something good and useful. The best and most useful thing they could foresee was to utilize nuclear energy for the peacetime generation of electricity.

(As an insight into the physicists' psyche, a plan to use nuclear explosives for monumental construction projects, such as digging a new Panama Canal, was actually named Project Plowshare.)

Many of the scientists and administrators of the old Manhattan Project helped to staff the new Atomic Energy Commission, immediately after World War II ended. The AEC's task was to oversee the development of nuclear technology for both the military and civilian markets: oversee, promote, regulate

— the AEC had all the responsibilities, and power, in its own hands.

The predictions for nuclear power in the civilian market were sky high, and corporations like General Electric and Westinghouse jumped onto the bandwagon: Nuclear power was going to make electricity dirt cheap, now and for all time.

In their enthusiasm, driven perhaps by the specter of Hiroshima's mushroom cloud, the scientists made wild claims of trouble-free nuclear energy for everyone. The AEC and the major corporations worked together smoothly to develop the reactors that would be the heart of nuclear power stations.

Enter Hyman Rickover and his quest for nuclear submarines. Against the entire armamentarium of the US Navy, Rickover won his battle to create a nuclear submarine missile force. In doing so, he inadvertently pushed the corporations into developing *civilian* nuclear reactors that were strongly influenced by the requirements for *naval* power plants.

To this day the nuclear reactors that American corporations deliver to the electric utilities owe much of their design philosophy to the Navy's requirements for submarine reactors. Other types of reactors, for example the very reliable Canadian CANDU system, simply have never been allowed to compete within the US.

Physicist Freeman Dyson, of the Institute for Advanced Study at Princeton, has a slightly different view of the situation. In the early 1950s he was part of a small team of brilliant men who worked in a little red schoolhouse (literally!) in San Diego on nuclear reactor designs.

Dyson wrote in *Disturbing the Universe* (Harper and Row, 1979):

> The fundamental problem of the nuclear power industry [today] is not reactor safety, not waste disposal, not the dangers of nuclear proliferation, real though these problems are. The fundamental problem of the industry is that nobody any longer has any fun building reactors . . . Sometime between 1960 and 1970, the fun went out of the business. The adventurers, the experimenters, the inventors,

were driven out, and the accountants and managers took control
... We are left with a very small number of reactor types in operation, each of them frozen into a huge bureaucratic organization that
makes any substantial change impossible, each of them in various
ways technically unsatisfactory, each of them less safe than many
possible alternative designs which have been discarded.

While the nuclear power industry was being born in the US,
similar infants were growing in Canada, Great Britain, France,
and the Soviet Union. West Germany, Japan, and several Eastern European nations entered the game somewhat later, as did
Israel, Iraq, Brazil, and Argentina. Today, nuclear power industries exist also in India, China, and several Arab nations, as well
as in South Africa.

Opposition to nuclear power was relatively slow in forming,
but today it is formidable and very well financed and publicized. The main problems that the Luddites see with nuclear
power are those that Dyson enumerated:

1. *Reactor safety.* The Luddites claim that nuclear power
stations are "unsafe at any speed." They fear radiation leaks.

2. *Waste disposal.* Radioactive wastes from reactors are accumulating and no plan for their safe disposal has been agreed
to. Some of these wastes will remain radioactive for thousands
of years.

3. *Proliferation.* Nuclear power industries give nations the
capability of making nuclear weapons. Proliferation of nuclear
weapons, especially among unstable governments or terrorist
groups, is tantamount to suicide for the human race.

As a resident of Connecticut, I have a personal interest in
nuclear power. During the bitterly cold winters of 1977–78 and
1978–79, when shortages of natural gas created desperate emer-
gencies in the Midwest, the New England states suffered no
shortages of heat or electricity, largely because more than half
the region's electricity is supplied by a half-dozen nuclear
power stations.

The entire electric utility industry of the US has a total capacity of about 515 gigawatts (515 thousand million watts). This amounts to the equivalent of burning roughly 10 million barrels of oil per day. Some 45 gigawatts of that total is nuclear: nearly nine percent. Every day nuclear power stations save us roughly 890,000 barrels of oil. Since we currently import about 47 percent of our oil, nuclear power stations save us some 418,300 barrels of imported oil daily. At a nominal price of $20 per barrel, that is $8,366,000 *per day* that does not go into Arab treasuries.

According to the bills I receive, Connecticut's electricity rates are lower, thanks to nuclear power, than they would be if the region had to generate all its electricity from oil, coal, or natural gas. That statement flies in the face of Luddite claims that nuclear power is too expensive to continue, I know. But while the capital cost of a nuclear power station is bound to be higher than the cost of building a fossil-fuel power station, the *operating* costs are lower because a nuclear power plant requires pounds of fuel where a fossil plant requires thousands of tons.

In the authoritative Report of the Energy Project at the Harvard Business School, *Energy Future* (Random House, 1979), edited by Robert Stobaugh and Daniel Yergin, the following statement is made:

> . . . almost six years after OPEC quadrupled the price of fossil fuels . . . it is still plausible to assert that atomic energy is or is not competitive [over fossil fuels] by a choice of assumptions that suits one's interest.

In other words, the experts cannot tell whether nuclear power is cheaper than fossil-fueled power or not. How can I say that it is? Simple. I get the bills. I have an apartment in Manhattan and a home in Connecticut, three hours' ride from New York. The Connecticut rates are lower, by far. The percentage of electricity generated by nukes is higher, by far, in Connecticut than it is in Manhattan.

I realize that Northeast Utilities has far fewer economic headaches than New York's Con Edison. But executives in both utility companies agree that nuclear power costs less than fossil fuels, and the bills they send me underscore that conclusion.

Still, the Luddites insist that nuclear power plants are not safe, that nuclear wastes are a lethal problem for millennia to come, and that "peaceful" nuclear power industries are cloaks for nuclear weapons development.

Are nuclear power plants dangerous? Yes. So are automobiles, cigarettes, and swimming pools.

If you want to think about a truly dangerous energy source, think about natural gas. When I was a kid, growing up in the row houses of South Philadelphia, the only time we got a playground was when a gas main exploded and took out a house or two. The city smoothed over the rubble and we had a place to play.

Even today, how many people die each year in fires caused by propane tank explosions?

We accept lung-rotting pollutants from coal and the explosive hazards of natural gas because they are old, familiar dangers; we have lived with them (and died with them) for generations. The big danger from nuclear energy is radiation, which is invisible and insidious: You don't really know if it's hurt you until twenty or thirty years downstream.

Worst of all, it's a *new* danger. It hasn't been around long enough for us to accept it as a natural part of the background to life, such as 50,000 deaths per year from highway accidents.

The old AEC attitude toward questions of nuclear safety was a lofty, "Go 'way boy, don't bother me." With the haughty assumption that the general public would neither understand nor care about technical matters, the AEC worked hard for decades at building public distrust.

Certainly their biggest mistake — one that was carried over into the reorganization that ended the AEC and eventually created the Department of Energy — was to put the Nuclear Regulatory Commission under the same roof as the rest of the boys. NRC's job is to police the nuclear industry, to look out for the public's safety. But when the policeman reports to the people who are pushing full-steam-ahead for more and bigger nuclear power stations, the policeman tends to get shuffled off to one side.

The promoters of nuclear power promised cheap clean energy and brushed aside questions of safety. By the 1970s, though, the environmental movement had gained enough credibility — and clout — to challenge the claims of the pronuke forces. And several members of the NRC, disgruntled and frustrated, broke ranks to reveal to the public their misgivings about the industry and the safety of nuclear reactors.

In the intervening years, much nonsense has been written and spoken by both sides of the debate. By that I mean nonsense. The graybeards from the old AEC and the new Department of Energy reacted to criticism with the typical bureaucrat's knee-jerk: "It's not true, and besides, we can fix it without any help from you guys. And no, you're not allowed to see the reports." The anti-nuke lobby easily adapted the passions and techniques that had been developed during the resistance to the folly of Vietnam. "Hell no, we won't go!" soon turned into, "No more nukes!"

I grew up writing and editing newspapers, and I recognize a slanted story when I see one. When I began to see scare headlines proclaiming, "WE ALMOST LOST DETROIT," and "MILLIONS THREATENED BY NUCLEAR MELTDOWN," I realized that the dam had burst. In most cases, the stories under those headlines related accidents no more serious than a valve jamming, or a carpenter dropping a hammer on his foot.

No one has ever been killed in a nuclear reactor accident. No member of the public has been hurt by a nuclear plant accident. Even in *The China Syndrome*, Jane Fonda's film about nuclear safety, the power station's computerized safety equipment shuts down the reactor quickly, easily, automatically, before any harm is done to anyone — except to Jack Lemmon, who is gunned down by the National Guard.

Is the glass half full or half empty? Many people saw in *The China Syndrome* a vivid warning of the dangers of nuclear power, compounded by the greed of corporations and their lackeys. I saw a power station's equipment function the way it was designed to do — maybe *because* Jack Lemmon was shot down and could no longer mess with the controls.

Which brings us to Three Mile Island. The accident that occurred in Unit 2 of the complex of nuclear power plants on Three Mile Island was caused, according to the government's report, by a combination of equipment failure, ambiguous instrument readings, and operator misjudgments. More and more, as investigators sift the evidence, it appears that human error was the basic cause of the accident.

On March 28, 1979, at almost precisely 4 A.M., the pumps that feed cooling water to Unit 2's nuclear reactor shut down. This is not unusual, and the emergency backup pumps automatically started up, as they are designed to do. But, unknown to the operators in the control room, the valves in these emergency pumps had been closed, probably several days earlier, and were never opened again as they should have been. So the emergency pumps could get no water and a sequence of events started that quickly led to a dangerously overheating nuclear reactor.

The operators in the control room aggravated the problem, rather than alleviated it, because they misread some of the instruments on their display panels. One of the worst things they did was to turn off the emergency system that cools the nuclear reactor's core after it had automatically turned on and started to do its job.

The result was that the core of the reactor nearly melted down. Some radioactive xenon 133 was released into the atmosphere. Xenon 133 has a half-life of about five days. Government scientists estimate that the amount of radioactivity released, approximately 4000 man-rems, might cause one cancer death over the next 30 to 40 years.

Not as dangerous as a carton of cigarettes, perhaps. Far less chancy than driving a hundred miles on a superhighway. But dangerous enough.

It now seems clear that there was never any chance of explosion from the famous "hydrogen bubble" that built up inside the reactor's pressure chamber. But the fear of an explosion that would rupture the chamber walls and pour lethal radioactivity all through the countryside caused a lasting trauma in the minds of every thinking person who heard about the accident.

Although the worst possibilities of disaster were averted, narrowly, that March weekend was still a very terrifying one for the citizens of Middleton and Harrisburg — and the world.

The political repercussions of the Three Mile Island accident are still reverberating around the globe. In the United States, critics of nuclear power claim that "the nukes are dead." In nations like Sweden, public referenda on nuclear power have been taken. While no nation has voted to stop nuclear power plants altogether, it is clear that public misgivings are vast and deep.

Still, other nations, for example, France, are going ahead with firm development schedules, despite public protests. André Giraud, Minister of Industry, has said, "We do not accept . . . the idea that an objection from one person or just a small number of persons would go against the opinion of the majority of the citizens . . . We do not accept the law of minority . . ."

It is fascinating to note that over the same weekend in which the Three Mile Island accident riveted everyone's attention to Pennsylvania, a gas line explosion in Massachusetts totally demolished a suburban shopping center, a trainload of toxic chemicals derailed in Florida and residents of a five-mile-wide area had to be removed to safety from the lethal fumes, and an elephant ran wild in a circus in the Midwest and trampled several people.

How safe is safe? In the final analysis, only public opinion can answer that question.

The Luddites see nuclear power as a classic example of technology run wild, with Big Business — aided and abetted by Big Government — riding roughshod over the people.

The Prometheans agree that the nuclear industry needs better-designed power plants and better-trained operators for those power plants. But they insist that nuclear power is safe, can be made even safer, and is a necessary part of our energy technology for the next generation or two. Instead of discarding nuclear power, the Prometheans believe that we can make it useful and acceptable by employing some of the tools and techniques that we have developed in the space program.

17

Nuclear Power: The Good News

People do not lack strength; they lack will.
—VICTOR HUGO

Even though they disagree on how serious the problems of nuclear power are, Luddites and Prometheans both see three main problem areas for nukes: reactor safety, waste disposal, and weapons proliferation.

What can space technology offer to help solve those problems?

Consider reactor safety first. The equipment of nuclear power stations is reliable in the extreme. Nuclear power plants have been designed with all the redundant emergency backup systems of manned spacecraft. These systems have proved that they work, and work well, time and again over many years, in tests and in actual operation.

Certainly some improvements can be made on detailed aspects of the equipment. But the biggest gain in safety stands to be made not by improving the equipment, but by improving the performance of the personnel who operate the equipment. Equipment is only as reliable as the men and women who operate it.

First is the matter of human engineering, or what scientists are now calling *ergonomics*. This is the science (art?) of designing equipment with the needs of its human operators in mind from the very first touch of pencil to graph paper.

"The design of [nuclear control] systems is simply in an intolerable state," according to Donald Norman, a psychologist at the University of California at San Diego who specializes in ergonomics.

In the control room of Three Mile Island's Unit 2, for example, many controls were located several steps away from the instruments that showed the operators the conditions they were supposed to be controlling. The operators had to run back and forth from instruments to controls to do their jobs. Many of the instruments were difficult to read, especially under emergency conditions. Some were in the glare of strong lights, while others were in dim light or even obscured by other equipment. Some were actually hidden from the operators' sight. On one valve control a red light meant that the valve was open; on a nearby valve control a red light meant it was closed.

With claxons sounding and nerves raw with tension, how could anyone expect the operators to work such a mishmash smoothly and effectively?

When NASA first began designing the manned spacecraft of the Mercury series, they encountered very similar problems. Once the astronauts looked at the earliest mockups of the control panels, they quickly saw that they would have to be contortionists to use the controls.

"No way," said the astronauts. And because they were centrally important to the space program, the engineers listened. And learned.

Ergonomics was born.

The astronauts themselves became part of the design team, with the result that throughout the one-man Mercury, two-man Gemini, and three-man Apollo programs, the instruments were grouped and placed where they could easily be seen, understood, and operated by the astronauts.

The control rooms of nuclear power stations should be designed with the same philosophy in mind from the beginning. We have the knowledge and the skilled people to accomplish such a task. The Nuclear Regulatory Commission must see to it that the utilities who own and operate nuclear power stations design and build their control rooms around ergonomically sound principles.

The other enormous contribution that space technology can make to nuclear safety lies in the training of the nuclear power plant operators. Space technology does not consist merely of hardware. It consists of people: trained, dedicated people who have developed management techniques, training methods, and a mental attitude best described by the old phrase *esprit de corps*.

Again, we can benefit from a brief look at the history of manned space-flight programs: Mercury, Gemini, and especially Apollo.

Back in the days of Vanguard, when American rockets were failing about as often as they successfully rose off the launch pad, many pundits claimed that rocket boosters were so inherently complicated that no one could ever make them work reliably. The doomsayers trotted out mathematical proof: There are so many thousands of working parts in even a rocket as simple as Vanguard that, statistically, the chances of their all performing properly at the same time, in the proper sequence, are vanishingly small. This problem, they said, would make it impossible to trust human lives to utterly unreliable rockets. Manned space flight was impossible, they claimed.

Somehow the Russians managed to get Yuri Gagarin into orbit in 1961. By 1963, six Russians had orbited, including Valentina Tereshkova, the only woman who has ever been in space — so far. By 1964 the Soviets were orbiting two-man spacecraft, and in March 1965 Alexei Leonov became the first human to leave his spacecraft and "walk" in space.

Obviously, the Russians knew how to beat the doomsayers' statistics.

And so did NASA. Because behind the numerology was the real problem of rocket reliability: quality control. Anyone who has bought an automobile knows that a new car requires a "breaking in" period, during which you drive the auto to see what's wrong with it: a squeak here, a faulty windshield wiper there, a sticky trunk lid, a pinging in one of the cylinders. Once I bought a new Dodge that developed a mysterious and annoying clunking noise. The dealer's mechanics finally found the cause: An empty Coca-Cola bottle had been left inside the paneling of the right front door.

In standard industrial practice, quality control is done on a batch basis. Out of every x number of parts, one is grabbed by the quality control inspector for testing. It is assumed that a certain number of defective parts leave the factory, but the number is presumably small and a tighter quality control check would cost more money than the manufacturer believes worthwhile.

NASA learned that when it comes to complicated rocket boosters — especially boosters intended to carry astronauts into space — the only permissible number of defects in the equipment is *zero*. So NASA started its Zero Defects Program. Everything was checked and checked and checked again. Everyone knew that Zero Defects was impossible, but by God every NASA technician had that goal plastered before his eyes no matter where he turned. Huntsville, Cape Canaveral, Houston, even in the headquarters offices in Washington, Zero Defects became a way of life.

Human lives depended on getting the defects in equipment down as close to zero as possible. And then working still harder to reduce the defects even more.

The Zero Defects Program was a mental attitude as much as anything else. *Esprit de corps.* When those astronauts went roaring off into space on tongues of bellowing flame, they knew that every switch, every transistor, every oxygen line was the product of the sweat and patience and skill and integrity of thousands of technicians who worked night and day to keep them alive.

This attitude of professionalism, this dedication and *esprit de corps*, carried over into the training and performance of the mission control engineers as well.

In April 1970, the Apollo 13 spacecraft carrying James Lovell, John Swigert, and Fred Haise to the Moon suffered a nearly catastrophic explosion in an oxygen tank. The mission controllers in Houston, backed by a vast team of NASA and industry engineers, worked unceasingly around the clock for days to figure out how to jury-rig the undamaged equipment aboard the crippled spacecraft to get the astronauts home safely.

They had to swing the spacecraft around the Moon. They had to use backup systems in ways they had never been intended to be used. They had to invent and improvise. And they did it. Lovell, Swigert, and Haise made it back to a safe splashdown, and the whole world cheered.

Compare the smoothly professional work of those controllers in Houston with the panicked behavior of the operators in Three Mile Island's Unit 2.

Training. Professionalism. Dedication. *Esprit de corps.*

This is what our hard-won knowledge of space technology can offer the nuclear power industry. The operators of nuclear power stations should be trained far beyond the levels of ordinary electric utility employees. Perhaps they should not even be trained by the utilities that own the power plants.

What we need is a Plutonium Priesthood* of trained and dedicated men and women who will operate nuclear power stations with the public's safety uppermost in their minds, and a Zero Defects attitude emblazoned in their hearts. A Plutonium Priesthood, recruited as volunteers from school, trained by experts from the space program, devoted to the

Plutonium Priesthood is a term I borrow from science fiction writer Arsen Darney, who used it in another context entirely.

public safety, and responsible to a specially created safety commission *and no one else,* can ensure us the best possible performance in the control rooms of nuclear power stations and the highest levels of maintenance of the equipment.

A Plutonium Priesthood will be expensive. So will a Zero Defects program. But if the nuclear power industry wants to regain the confidence of a skeptical American public, such a priesthood and such a program are necessary — and far cheaper than the billion-dollar cost of cleaning up Three Mile Island.

No matter how safely nuclear power stations are operated, nuclear reactors produce radioactive wastes. Spent fuel rods have been accumulating at nuclear power sites for decades while the government dithers over the problems of where to dispose of the radioactive wastes.

Opponents of nuclear power remind us constantly that these wastes will be dangerously radioactive for thousands of years. Environmentalists insist that there is no truly safe way to dispose of them. Wherever they are stored there is a chance that an earthquake, or erosion, or some other unforeseen event will crack open the containers and spew deadly radioactivity upon us.

After all, they claim, these wastes must be stored for millennia and nothing can be guaranteed for that long.

Wrong.

In less than twenty years it should be perfectly feasible to boost nuclear wastes off the Earth. They can be "parked" in orbits deep in space or landed on the Moon. They can be dropped into the Sun or hurled out of the Solar System forever.

The choice of where to put the radioactive wastes depends mainly on how much rocket fuel and money we are willing to spend on the task. The desirability of lifting them off-planet seems undeniable.

A typical 1000-megawatt nuclear power station will generate some four tons of radioactive wastes per year. Compared with the 500-plus tons of soot and sulfur dioxide emitted by a coal-fired 1000-megawatt plant *per day*, this does not seem like much of a problem.

The wastes from a nuclear plant are radioactive, of course. But many of the waste products are low-level radioactives, with half-lives of less than a year. Iodine 131, for example, has a half-life of 8.05 days. Curium 242's half-life is 163 days.

These low-level radioactives could most likely be stored on Earth safely and allowed to decay into inert elements. But the deadlier radioactives, such as Strontium 90 (half-life of 28 years) and Plutonium 239 (24,360 years) could be packed aboard rocket boosters and hurled off this planet forever. Getting rid of the plutonium, by the way, also alleviates the problem of weapons proliferation. It is the plutonium "waste" from commercial power plants that Luddites fear would be used for clandestine bomb making.

Shades of Skylab! Can we trust deadly radioactive wastes to rocket boosters? What if the rocket blows up on the launching pad, or the orbiting ashcan spins back into the atmosphere? If we design the boosters and the waste containers with the same skill and care that was used in designing the astronaut-carrying Saturn V boosters and Apollo spacecraft, we should be able to handle the problem safely.

Certainly we would boost the wastes into very deep orbits, where there would be no possibility of their re-entering the Earth's atmosphere. It may be desirable to soft-land the wastes on the Moon, or to park them in orbits around the Moon. If we wish to spend the money and energy, they can be flung into the Sun or entirely out of the Solar System, as are our Pioneer and Voyager probes.

The containers for radioactive wastes must be crashproof, designed to withstand the crash of a booster without cracking open. They would also have to be heavily lined with radiation shielding, so the crashproofing should not add an untoward amount to the total payload package's weight.

The waste containers would also have their own emergency escape systems, just as the Apollo spacecraft did. If the booster should malfunction during launch, the payload section automatically jettisons itself away on its own escape rockets and parachutes softly back to Earth for recovery and refit onto another booster.

By the year 2000, the US alone could be producing 500,000 tons of radioactive wastes per year. While not all of this needs to be lifted off-planet, it would take almost daily launches of a Space Shuttle to handle the load. On the other hand, by A.D. 2000, updated versions of the Shuttle should be flying, with increased payload capabilities. Or simpler throwaway boosters might be the cheapest way to handle the radioactive waste problem.

The cost of boosting radioactive wastes off-planet will probably amount to a small fraction of the cost of operating the nuclear power plants that produce the wastes. This is far from an insignificant amount of money: It adds up to billions of dollars per year; but it is not so expensive, given the size of the industry, that it would be economically prohibitive.

What about the proliferation of nuclear weapons?

Michel Hug, of Electricité de France, says flatly, "Leading people to believe that . . . controlling civil uses of atomic energy will prevent proliferation is a pure lie, because there are several ways which are cheaper for a country to get nuclear weapons than to have an energy program."

Any nation that wants to develop a nuclear arsenal, and can spend the money to do so, can become a nuclear power. It is not necessary to develop a nuclear power industry first, or at all.

The "secret" of nuclear power was revealed to the world at Hiroshima. A bomb that worked could be built. All the spies and later intrigues dealt with the details of how to build such weapons. The physics was clear, even simple. The engineering was difficult, but if one nation could do it, so could others. And they did.

If the United States foreswore nuclear power altogether and shut down every nuclear station in the land, it would not prevent other nations from building nuclear weapons. If every nation in the world shut down its nuclear power plants, it would not prevent anyone who wanted to from developing a nuclear strike force. It might make the project more difficult, more expensive — but certainly not impossible.

There are two contributions that space technology can make toward controlling nuclear weapons proliferation.

One already exists: Vela satellites orbit our planet, on watch against nuclear bomb tests on the surface of Earth or in space. Several times, such satellites have spotted bursts of energy in deep space and reported the data automatically to control stations on Earth. In each case, the energy bursts turned out to be natural astronomical phenomena.

In 1979, however, a Vela satellite reported a suspicious burst on Earth's surface, near South Africa. No official confirmation of the source of that burst was ever made, although international rumor had it that it was a clandestine test of an Israeli nuclear weapon.

We have the technology to spot nuclear weapons tests, even tests conducted deep underground. What we lack is the international legal system to prevent nations from developing nuclear weapons. Or using them. We may have, though, the means for shooting down bomb-carrying missiles, within the next decade. As we will see in Chapter 25, satellites armed with energy-beam weapons may be patrolling the skies in the 1990s. They could make an effective antimissile system that would prevent ICBMs from reaching their targets.

Such space-born weaponry, however, would not stop other delivery systems. While a full-scale nuclear Armageddon may be averted by satellite defenses, terrorist organizations could still acquire nuclear weapons and smuggle them into any city on Earth.

Certainly a group as large and well-financed as the Palestinian Liberation Organization has the capability to acquire or even develop nuclear weapons. Student radicals working in a university basement — probably not. Chances are they would kill themselves with radioactivity or chemical poisoning (plutonium is an incredibly vicious poison) long before they had assembled a workable bomb. But the prospects are certainly chilling.

As the gray old *New York Times* put it editorially on December 23, 1979:

> Safety will make or break nuclear power in this country. The proliferation of weapons abroad will not be affected by domestic power plants . . . Nuclear power is too valuable to reject without very good reason.

We can use the strengths we have gained from space technology to increase enormously the safety of nuclear power plant operations and to remove forever the problems of nuclear waste disposal. We may even be able to use space technology to protect ourselves against nuclear attack. Will we do so? Even if we don't, perhaps others will. As France's Michel Hug said:

> Do you know people who accept today to work the way their father, or grandfather, did? . . . Do you know people who would accept to push wheelbarrows instead of driving trucks? I don't know so many of those people. And those ecologists, they are not ready to push wheelbarrows, either.

18

Hydrogen Fuel

> Water, water, every where,
> Nor any drop to drink.
> —SAMUEL TAYLOR COLERIDGE

The largest NO SMOKING signs in the world were painted on two enormous tanks that held liquid hydrogen at Kennedy Space Center's Launch Complex No. 39, where the Apollo rockets lifted off for the Moon.

The upper stages of the mammoth Saturn V booster were fueled by hydrogen, which when burned with oxygen makes not only one of the most energetic propulsion systems known, but one of the cleanest. The two tanks, each two hundred feet in diameter, held 476,000 pounds of liquid hydrogen apiece.

We went to the Moon on hydrogen fuel. If we use the knowledge gained from space technology, we can replace our dwindling supplies of petroleum with hydrogen. Not only will hydrogen be cheaper in the long run; it will be far, far cleaner. A hydrogen-burning engine emits water, not carbon dioxide, or carbon monoxide, or sulfur oxides: just hydrogen oxide — plain old water.

The late Shah of Iran once said that oil is much too valuable to burn. He merely echoed the knowledge of petrochemical engineers. Petroleum is the basic feedstock of the entire plastics industry. Oil products are also basic to the manufacture of fertilizers, cosmetics, lubricants, and thousands of other products, large and small. To burn oil for transportation or heating is criminally wasteful.

It took nature some 200 million years to create petroleum. We will have exhausted the world's supply in not much more than 200 years. For despite the urgings of the oil companies, there is little more oil to be found on planet Earth. Since the boom in drilling for new oil wells started with the Arab Oil Embargo of 1974, no new major oil reserves have been found.

World oil reserves are decreasing; they peaked in the mid-1970s at roughly 675 billion barrels. By 1979 they were down to 652 billion barrels, and falling. Meanwhile, global demand for oil increased at three percent per year during the same period.

Lewis Beman, of *Business Week* magazine, wrote in *The Decline of U.S. Power* (Houghton Mifflin Company, 1980):

... the trend is clear and alarming: for the first time since oil became a major source of energy, the world's factories, fleets of cars, and electricity plants are burning up oil faster than it is being discovered.

It is certain that we must find an alternative to petroleum for those "factories, fleets of cars, and electricity plants." Solar energy alone cannot do the job. Hydrogen could.

If oil is too valuable to burn, is water too valuable to drink? Hydrogen comes from water. Fortunately, our planet is brimful of water. And when you burn hydrogen in air or pure oxygen the combustion product is water; hydrogen fuel recycles immediately. You don't have to wait 200 million years for another crop of dinosaurs, or even 200 seconds. Burn hydrogen and you get water vapor coming out of the exhaust pipe.

Hydrogen can be used wherever fossil fuels are used today: in the engines of automobiles, aircraft, trains; in heating systems for homes and other buildings; in electric power plants that today burn oil, coal, or natural gas. And since hydrogen burns so cleanly and efficiently, virtually all the air pollution coming from fossil-fuel burning will disappear once hydrogen goes into widespread use.

If hydrogen is that good, why isn't it being pushed hell-for-leather by everyone? Two reasons: cost and safety. Maybe a third and fourth reason, as well: ignorance and inertia.

Safety first.

Wherever the idea of hydrogen fuel is proposed, someone is bound to frown with puzzlement for a moment, and then burst out, "My God, the *Hindenburg!*" The 1937 disaster in which the hydrogen-filled German dirigible *Hindenburg* exploded and burned before a horrified crowd at Lakehurst, New Jersey, has done more to delay the serious development of hydrogen fuel than all the protesters have done to delay the development of nuclear power.

The *Hindenburg* was filled with hydrogen, which is the lightest element in the universe, and therefore the best gas for a lighter-than-air dirigible. But the stuff burns easily, and with a very hot flame.

No one knows what set off the *Hindenburg*'s hydrogen. The best guess is that it was a spark of static electricity. There had been thunderstorms in the Lakehurst area earlier that day. In any event, the hydrogen exploded. The dirigible burned to a twisted heap of aluminum girders within minutes, and in those minutes the transatlantic air transport line that had been started by Count von Zeppelin decades earlier ended forever. Thirty-five passengers and crew members died in the flames. Sixty-two survived. Many people, to this day, are surprised to learn that anyone survived.

One of the most famous news photographs of all time caught the *Hindenburg* half engulfed in the flames. Try to visualize that photograph in your mind. What are the flames doing? They are shooting straight up, because hydrogen is so light that a hydrogen fire does not spread sideways, as a gasoline fire does. It rises straight up.

Remember that We will get back to it.

Engineers who want to develop hydrogen as a fuel recognize the *Hindenburg* syndrome, and as a result they have formed the *H₂indenburg* Society. It's a refreshingly uncomplicated club, with no officers, no meetings, no newsletters, and no dues. It is a loose, very informal association of men and women whose one link is that they believe hydrogen should be developed as a fuel to replace oil, natural gas, and even coal.

The only tangible evidence of the *H₂indenburg Society* is a button, pure white, with the word *H₂indenburg* in black German script across it. The H_2 is the chemical symbol for molecular hydrogen, which is the form of hydrogen that is used as a fuel: the H_2 in H_2O, water. The true purpose of the Society, of course, is to combat the ignorance and inertia that are holding back the development of hydrogen as a fuel.

As a member in good standing of the Society, I once asked the president of Atlantic Richfield Corporation what his company was doing about hydrogen fuel. He had just finished a speech in which he assured the audience that Atlantic Richfield was bending every effort to solve the energy crisis. The gentleman looked bemused for a moment, then replied carefully that his engineers had looked into the problem, but hydrogen is very dangerous, you see, and it's much better to drill for new oil wells.

At the same time that we were chatting, the Daimler Benz Corporation (the folks who make the Mercedes automobile) were starting up a fleet of hydrogen-powered buses in Stuttgart.

Hydrogen *can* be very dangerous stuff. A very energetic fuel, it burns at a high temperature. In its gaseous phase it is quite volatile; it explodes easily. Yet hydrogen-oxygen rockets carried our Apollo astronauts to the Moon and back quite safely. The only problem during the entire program was an explosion of an oxygen tank aboard Apollo 13. The hydrogen behaved perfectly.

The hydrogen used in Apollo and the Saturn V rockets was liquefied, cooled down to a cryogenic temperature of $-423°$ F.

That is —252.8° C, only 20.4 degrees above Absolute Zero.

It took NASA's engineers a lot of years and many spectacular failures to learn how to produce, handle, and use liquid hydrogen. But that knowledge has been won, and we now routinely produce the cryogenic temperatures needed for liquefying hydrogen, oxygen, and many other gases — even helium, which has the lowest boiling point of them all, only 4.3 degrees above Absolute Zero.

But no sane engineer suggests using liquid hydrogen to power the family automobile. Daimler Benz and other teams of researchers have hit upon a simpler, cheaper, safer idea. Hydrogen can be stored at normal temperatures, quite safely, if it is linked chemically to such metals as iron or titanium. Such chemical combinations are called *hydrides.*

The buses that Daimler Benz operates have fuel tanks filled with metal chips. Hydrogen gas is pumped into the tanks under pressure, and immediately forms hydrides with the metal. To coax the hydrogen free of the chips, a small heating element bakes some of the hydrogen free, and it is then piped into the engine for start-up. Once the engine is running, its own heat is fed back to the fuel tank to bank off more of the hydrogen fuel.

Hydrogen fuel tanks must be much larger than gasoline fuel tanks, because hydrogen is much bulkier than gasoline even though it is lighter. But a tank filled with metal-chip hydrides could be machine-gunned without bursting into flames.

Hydrogen offers other safety advantages over such petroleum fuels as gasoline. When hydrogen burns, the flames flow straight up. Gasoline tends to flow along the ground, often surrounding a crashed car in a pool of flaming fuel.

Many tests have been made to show hydrogen's safety. The major stumbling block concerning safety at the moment is to convince the nation's decision-makers in government and industry to *look at the available information.*

Today's automobile engines can be converted to run on hydrogen. In fact, a few years ago a precocious teen-ager named

Roger E. Billings converted a Model A Ford that was still in the family to a hydrogen-powered car — despite his father's doubts. Billings is now the head of the Billings Energy Corporation, which is delivering hydrogen technology for transportation, heating, power, and even converting a whole midwestern city to hydrogen fuel.

The other side of the safety coin is environmental safety. Here hydrogen stands virtually alone. Fossil fuels are inherently dirty, because they are composed mainly of carbon and hydrogen — hydrocarbons — plus traces of sulfur and other elements. Inevitably, fossil fuels pour out pollutants such as oxides of nitrogen and sulfur, which cause smog, poisonous carbon monoxide, and carbon dioxide, which adds to the worldwide greenhouse effect.

Hydrogen engines emit some oxides of nitrogen, simply because there is nitrogen in the air that is burned with the hydrogen. But there are no sulfur oxides and no carbon monoxide or carbon dioxide, because there is no sulfur or carbon in hydrogen fuel. The bulk of the exhaust is simply water vapor.

It takes plenty of energy to tear the hydrogen away from the water molecule, and the cost of this energy is what makes hydrogen fuel so expensive. At present.

Hydrogen enthusiasts foresee the day when hydrogen is produced on such a vast scale that it will be economically competitive with petroleum. Certainly, as oil gets more scarce and more expensive, and hydrogen technology improves, there will be a crossover point. If current trends continue through the 1980s, hydrogen will be economically competitive with gasoline in the 1990s.

Moreover, the scientists and engineers have a few tricks up their sleeves that may bring the cost of hydrogen down to earth much sooner. Researchers at several universities are experimenting with the possibilities of using sunlight to split water into hydrogen and oxygen. Michael Gräetzel, of the École Polytechnique Fédérale in Switzerland, has developed a rubidium-based catalyst that splits water with no other energy input than sunlight. While such experiments are far from commercially reliable practice, they point one possible way to very cheap hydrogen fuel. If we can extract hydrogen from nothing more than sunlight and water (plus a catalyst that, though expensive, is not consumed by the process of water-splitting), then the end of petroleum as a fuel is in sight. So is the end of the energy crisis, as well as most of the world's air pollution.

Even if that prospect does not pan out, the hydrogen enthusiasts see a future Hydrogen Economy in which hydrogen is split from water through electrolysis, just as high school students do it in their chemistry labs, but on a vastly larger and more economical scale. Electricity can split water into hydrogen and oxygen. Plans for the Hydrogen Economy call for large electric power stations (nuclear or fossil fueled) to be used entirely to provide electricity for water electrolysis. Existing power stations could be used, and new ones built next to the sea or large inland water sources, like lakes or major rivers.

There is no reason why solar-electric or windmill power plants could not be used; anything that provides sufficient electricity will do. One plan calls for a long line of windmills erected on artificial islands off the coast, feeding hydrogen electrolyzed from seawater to the mainland through flexible pipes laid along the ocean bottom.

These power stations would run at a constant level, 24 hours a day, seven days a week, to provide the electricity that splits the water molecule. No peak loads, no seasonal demands. Instead of providing electricity for consumers, these power stations will provide hydrogen for consumers. Thus they can run much more efficiently than power stations that must be scaled to meet peak demands, even though they run at much lower levels for most of the time.

The hydrogen produced by electrolysis can then be piped cross-country through existing pipelines originally laid down for natural gas, to the distribution points around the nation. There the hydrogen could be piped to homes for heating, to power stations for generating electricity, to "gas" stations for fueling autos and trucks and buses, to airports and train depots.

Each private home could become its own distribution center, taking in hydrogen to power the home's heater, drive its own electricity generator, and fuel the family car(s).

The Luddites see this as centralization, and therefore evil. The Prometheans see it as high technology, and therefore good. What would the Luddites say if and when equipment becomes available so that each householder can generate his or her own hydrogen directly from water? (And either sell the unwanted oxygen or use it instead of chlorine to clean the family swimming pool.)

The success of the Hydrogen Economy will depend on hydrogen's safety and price, rather than ideological considerations.

There's an old joke in the aerospace industry about a young woman who has just moved into town and goes to a local gynecologist for a checkup. The doctor examines her and, when reviewing the results of the examination with her afterward, keeps referring to her as "Miss Jones."

She informs him that she is "Mrs. Jones." The doctor is perplexed, especially after she reveals that she has been married for several years.

"But my examination showed that you are still a virgin," the doctor blurts out.

Mrs. Jones smiles, a bit sadly, and explains that her husband is an aerospace engineer. "Every night he sits on the edge of the bed and tells me how good it's going to be."

How good will hydrogen fuel be? Is it all an aerospace engineer's pipedream, too dangerous and too expensive for the practical world?

Young Roger Billings, who formed his own company, is now doing more than $8 million in sales per year. Not much, as major industry goes. But a good beginning.

One of Billings Energy Company's biggest projects is to turn the town of Forest City, Iowa, into a totally hydrogen-fueled community. Billings is designing a coal gasification plant for the town. Hydrogen released from the coal will then heat the town's public buildings and provide fuel for the town's industries, homes, and private automobiles.

Billings claims that the coal-gasification process can produce hydrogen for the equivalent of gasoline at 50 cents per gallon.

In 1980, President Carter proposed the creation of the Energy Security Corporation, which would direct an $88 billion program aimed at developing synthetic fuels, hydrocarbon fuels derived from coal. The technology is hardly new: The Germans developed it in World War I and depended heavily on it during World War II.

The synfuels program is championed by the major oil corporations, because they see synthetic hydrocarbon fuels as a replacement for today's petroleum. But synfuels will be more expensive than petroleum — estimates are that synthetic fuel from coal will cost at least $30 per barrel. And synfuel processes require enormous amounts of water, at least four times as much water as they produce in fuel. In the regions of the West where the coal is, water is scarce. A multibillion-dollar synfuels effort will put incredible strain on the water resources of the western states.

The same coal that is now proposed as the feedstock can be used to produce hydrogen, more cheaply than synfuels. And while producing hydrogen will require almost as much water as synfuel production, the water is returned to the environment as soon as the hydrogen is burned.

In the eyes of Billings and most other hydrogen proponents, the synfuels program is old technology, too expensive, too damaging ecologically, and far less desirable than a strong effort to produce hydrogen fuel.

Will American industry turn to hydrogen as swiftly as it can, or will inertia and ignorance keep us chained to fossil fuels, synfuels, foreign oil — until we are forced to buy hydrogen technology from West Germany or elsewhere?

Here again the Luddites, by resisting any move toward new technology, play into the hands of the very forces they profess to be against. Resisting new technology does not lead to a decentralized, solar-powered America. Resisting the development of hydrogen fuel pushes the nation into a wasteful and ecologically unsound multibillion-dollar synthetic fuel program, funded by government money and operated by the major oil corporations.

19

God's Chosen Energy Source

> Full many a glorious morning have I seen
> Flatter the mountain-tops with sovereign
>   eye,
> Kissing with golden face the meadows
>   green,
> Gilding pale streams with heavenly al-
>   chemy.
>
> —WILLIAM SHAKESPEARE

There is another form of energy locked inside the hydrogen atom which, even though it does not stem from space technology, must be included in any discussion of future energy plans.

It is nuclear fusion.

That golden sun that Shakespeare refers to, that the solar energy enthusiasts see as our salvation, is in fact a nuclear fusion generator.

Thermonuclear fusion. It makes the Sun shine. It powers the stars we see at night. You could say that it is God's chosen energy source, because the entire universe runs on it. Fusion promises virtually unlimited energy. Its fuel is water. Its pollution products are virtually nil. It produces very little radioactive waste. It could provide us with cheap, clean, abundant, high-grade intensive energy that could even be decentralized, if we so desired.

Fusion is the ultimate Promethean dream: to capture the heart of a star and harness all that titanic energy to do our bidding. Fusion is so promising that physicists around the world have spent billions of dollars and more than thirty years trying to achieve it. In the 1950s they believed they were twenty years away from demonstrating a successful fusion reaction in the laboratory. Today they believe they may be twenty years away from a demonstration power plant.

The public, seeing men in lab coats making tight-lipped predictions centered on a time two decades distant, tends to place fusion in the same category as black holes, flying saucers, and other incomprehensibilities.

But fusion power stations will be with us before today's infants grow to maturity. The question is, Will those plants be as good as advertised? In the 1950s we learned how to make uncontrolled fusion reactions: hydrogen bomb explosions. The game now is to produce controlled fusion reactors, to tap the star-hot plasmas for useful energy.

Although fusion energy comes from the heart of the atomic nucleus, it is a very different sort of process from the fission type of nuclear energy that we already use in electric power stations. In fission, heavy atoms of uranium or plutonium are split apart (hence the term *fission*) to yield energy. This process produces radioactive wastes. In fusion, light atoms such as those of hydrogen isotopes are forced together *(fused)* to release energy. There are no direct radioactive waste products, although the fusion reactor itself becomes highly radioactive in time.

It's really easy to make a controlled thermonuclear fusion reactor. All you need is some ten billion billion billion tons of hydrogen (10^{27} tons) and a million years or so. Let nature take its course, and at the end of that time you have a million-mile-wide star like our Sun: a very stable, long-lived fusion reactor.

Making a controlled thermonuclear reactor (CTR) on Earth is a bit more difficult. Physicists must work with teaspoonfuls of material, not billions of tons. This means that they cannot duplicate the conditions at the core of the Sun, where the solar fusion reactions take place. They must be clever enough to find a different set of conditions that they *can* produce in a terrestrial laboratory and that still yields a sustainable fusion reaction.

It appears to be impossible to produce fusion reactions on Earth with normal hydrogen atoms. So the physicists have turned to hydrogen's heavier isotopes, deuterium and tritium.

Deuterium exists in ordinary water. For every six thousand atoms of normal hydrogen atoms in water, more or less, there is one atom of deuterium. Tritium is practically nonexistent on our planet and is usually created in the laboratory.

Deuterium can be separated from ordinary water rather simply and cheaply, using conventional equipment. Each deuterium nucleus — called a deuteron — is twice the weight of an ordinary hydrogen nucleus, because where hydrogen nuclei are nothing more than single protons, deuterons consist of one proton plus one neutron.

One atom out of every six thousand hydrogen atoms in the water you drink, swim in, bathe with, is deuterium. And the fusion process is so rich in energy that one cubic meter of water (about 265 gallons) can yield more than 400,000 kilowatt hours of energy. From that one deuterium among the six thousand hydrogens. This means that an eight-ounce glass of water has the energy content of 500,000 barrels of oil. A sugar cube's worth of water contains the energy equivalent of 2000 barrels of oil.

One cubic mile of seawater can provide as much energy as all the known oil reserves on Earth. There are more than 316 million cubic miles of seawater in the Pacific, Atlantic, Indian, and Arctic oceans.

And that's using only one six-thousandth of the hydrogen in the water.

To put it on a more personal level, every time I flush the toilet, roughly 15,000 kilowatt hours of deuterium returns to the sea. That's enough to run my apartment in Manhattan for more than eleven years. One flush of the toilet.

There is enough deuterium in the waters of Earth to provide thousands of times more energy than the world now uses, for millions — if not billions — of years into the future; using less than two-thousandths of one percent of the water. In essence, all the water is still there for other uses.

In the words of the old song, "Nice work, if you can get it."

Producing conditions that are in many ways more stringent than the environment at the heart of a star has not been easy. The deuterium gas must be heated to millions of degrees. At such a fierce temperature, the electrons surrounding the atomic nuclei are stripped away, leaving bare nuclei in a sea of free electrons. This state is called a plasma by physicists, an ionized gas, the fourth state of matter (solids, liquids, gases, plasmas).

The deuterium plasma must stay together without dissipating for a time long enough to allow fusion reactions to start. And once started, the fusion reactions must be self-sustaining, continuous. The best physicists in the world have been unable to achieve this, in more than thirty years of trying. To date the plasmas have been smarter than the plasma physicists. But the physicists have learned much, and believe now that they understand how to build reactors that will indeed produce sustained fusion. Their main line of approach has been to build large reactors in which the star-hot plasma is manipulated, heated, and held together within powerful magnetic fields. An ordinary gas is not affected by magnetic fields, but a plasma is an ionized gas, which conducts electricity and can be moved, shaped, excited by magnetic energy.

The reactors, with their "magnetic bottles" for holding the high-temperature plasmas, have been given whimsical names over the years: Perhapsatron, Stellarator, Phoenix, Baseball. The one that works best, and is being copied all over the world, is a Russian machine: Tokamak. The name was coined from the Russian words for Torroidal Magnetic Chamber.

While the magnetic-containment approach has been the fusion physicists' main line, in the 1970s a new and promising approach sprang up: laser fusion. Instead of trying to hold the heated plasma together inside a magnetic field, in the laser approach the deuterium is packed into a tiny, almost microscopically small, plastic sphere. The sphere is then bombarded on all sides by beams from energetic lasers. The laser beams squeeze the deuterium-containing pellet and in a millionth of a second or less the pellet implodes. The temperature and density at the core of the implosion shoot up to conditions similar to those at the core of a star, and fusion reactions take place. Deuterium nuclei fuse together to form helium, releasing very energetic neutrons and heat energy.

Laser fusion works. It has produced tiny puffs of fusion. In a practical fusion power plant the pellets would be fed in a continuous stream to the focus of the laser beams, and energy would come out in a machinegun-like staccato.

Laser fusion has obvious military implications. At present, hydrogen bombs are triggered by fission bombs built of uranium or plutonium, to produce the fiery temperature needed to set off the fusion process. As a result, H-bombs produce radioactive debris, just as fission bombs do. A hydrogen bomb triggered by the laser technique would be a "clean" bomb. For whatever that's worth.

In general, fusion is much cleaner than fission. There are no radioactive waste products from the fusion power process. The by-product of fusion is inert helium gas, the stuff we use to fill children's balloons.

Inside the fusion reactor, where very "hot" neutrons are shot off by the fusion process, radioactivity levels will be so high that the reactors themselves may have useful lifetimes of only a few decades. But those neutrons are so energetic that they represent a valuable energy source and will be used as such, not allowed to escape outside the reactor.

For example, KMS Fusion, Inc., where the first successful laser fusion took place, is studying the possibility of using fast neutrons from the fusion process to split water into hydrogen and oxygen. Oxygen is already a valuable industrial product, and hydrogen will be a major source of fuel, replacing petroleum.

The neutrons can also transmute the element lithium into the hydrogen isotope, tritium. Although tritium is vanishingly rare in nature (on Earth), it makes an even better fuel for fusion reactors than deuterium. It may be possible to wrap fusion reactors in a "blanket" of liquid lithium, which will absorb the deadly neutrons and thereby be transformed into tritium fuel for the reactor itself.

Fusion neutrons may also be put to use to solve the major problem of today's *fission* power plants: the disposal of radioactive wastes. It may be possible to bombard the radioactive wastes with fusion-generated neutrons and transmute them into stable, nonradioactive elements. Like Merlin's magic wand, those neutrons may be able to transform a dangerous dragon into something harmless.

And while fusion may not need space technology to become successful, practical fusion power plants will have tremendous *applications* in space technology.

As physicist Robert Bussard has pointed out, a fusion-powered rocket would allow spacecraft to flit across the Solar System in days, rather than years. Go to Mars for the weekend. Travel out to the Asteroid Belt for a few weeks of prospecting and mining, then haul back several tons of precious metals for sale on Earth.

More important, if we need rare isotopes of hydrogen and helium to make more efficient fusion power plants, it may be possible to scoop these isotopes out of the deep, cloud-streaked atmosphere of the planet Jupiter, using fusion-powered engines to propel our spacecraft. Jupiter is rich in hydrogen and helium, and isotopes that are rare or even nonexistent on Earth are undoubtedly abundant in the massive atmosphere of the largest planet.

Critics of fusion say that it is a long shot, it will be expensive (and centralized), and it won't be practical until our grandchildren's time. But the enormous potentialities for fusion — for clean, abundant, and ultimately cheap energy — mean that it is a long shot well worth backing.

In 1980 the federal government passed the Magnetic Fusion Development Bill, which promises $20 billion in funding for fusion over the next twenty years. But the passage of this bill does not mean that the Congress will appropriate the funds, year by year, that the program calls for. Indeed, in the lopsided Republican victory of 1980, the bill's "father," Representative Mike McCormack (D., Wash.), lost his seat in the House.

If fusion never becomes practical, the average American taxpayer will have lost a few pennies per year; far less than we pay for entertainment or electricity. If fusion does work, the world will be transformed. We may be able to struggle through the next two or three decades with interim, stopgap energy technologies such as coal and uranium, and then see a new era of energy abundance where individual, decentralized energy needs are provided for by compact fusion reactors — human-made suns that will end our energy woes forever.

By taming the heart of a star we will be able to soar beyond the Solar System and visit other stars. Through fusion we can give the gift of abundance to our children, and the gift of freedom that goes with abundance. They will find it very puzzling that the history books claim there was an energy crisis in the twentieth century.

20

Solar Power Satellites

> You see the little rift? "Believe this, not be-
> cause it is true, but for some other reason."
> That's the game.
> —C. S. LEWIS

On any cloud-free night, from any spot on Earth where humans make their homes, you will be able to see it: a star burning brighter than all the rest, unwinking, shaming Sirius and Jupiter and even Venus.

Built by human beings.

On the satellite itself, its sheer colossal magnitude will take your breath away. More than ten miles long, three miles wide, a glittering spiderwork of metal structure and dark-gleaming solar panels. Hundreds of people at work, scattered over it like tiny ants, men indistinguishable from women inside their hard suits and bubble helmets. Floating weightlessly inside your own space suit, you stare outward along a miles-wide arm of the Solar Power Satellite and see nothing but stars flecking the blackness of infinity. Turning, the overwhelming majesty of the blue-white planet we call Earth drowns your senses with beauty: the home of humankind, pure and clean and beckoning.

That science-fiction scenario may be material for factual stories in magazines and on television before this century ends. But if even one Solar Power Satellite is built by then, it will be only after a titanic struggle. For the SPS is fast becoming the major battleground between the Luddites and the Prometheans. On the one side, Prometheans are already proclaiming the Solar Power Satellite as the salvation of the human race, capable of providing endless prodigies of electric power cleanly and economically. On the other, the Luddites shudder at the prospect of gigawatts-worth of microwave radiation beamed through Earth's atmosphere from an SPS to receiving antennas on the ground. Calling the entire idea monstrous, they insist that it will be more expensive and less useful than nuclear power. In between these conflicting forces is the concept of the Solar Power Satellite itself, invented in 1968 by Peter E. Glaser, a vice president of the Massachusetts-based firm of Arthur D. Little, Inc.

In concept, the SPS idea is elegantly simple. The Sun shines all the time in space. Place a satellite in a high orbit, where it is always in sunlight. Geosynchronous orbit will do nicely: 22,300 miles above the Equator. At that altitude and position, the satellite rotates around the Earth once in 24 hours, at exactly the same rate as the Earth itself rotates. Therefore the satellite appears to remain stationary over one fixed point on the Equator. (Arthur C. Clarke first pointed out the usefulness of geosynchronous orbits for communications satellites in 1946. There is a small but determined group of loyalists who are trying to get such orbits called Clarke Orbits.) At geosynchronous altitude, the satellite is exposed to five to ten times more solar energy than the sunniest regions of Earth's surface. And it is always in sunlight, except for about 72 minutes per year, at the time of the spring and autumn equinoxes, when it slips briefly into Earth's extended shadow. This means that the "storage problem" simply does not exist for SPS. Ground-based solar energy systems must be designed around the fact that for at least half the time, the Sun is not available. Nightfall and poor weather conditions mean that either ground-based solar energy equipment must have some sort of storage capacity, such as pools of hot water or electric batteries, or the user must have an alternate energy system to employ during the hours when the Sun is not on hand.

The SPS converts the copious and unfailing solar energy in space into electricity by using the well-known and reliable technology of solar cells, the kind of solarvoltaic technology that has been providing electricity for spacecraft since the first Vanguard satellite went into orbit in 1958.

That's the easy part. To get this electrical energy down to Earth, where we want it, means that the energy must be converted into a form that can be transmitted over more than 22,000 miles of space and a few hundred miles of Earth's cloud-laced, turbulent atmosphere, to some kind of receiver on Earth's surface.

Glaser originally suggested using microwaves to transmit the energy. Although later investigators have brought up the idea of using laser beams, microwaves still seem to be the most likely means of beaming the energy from the SPS to the ground. Beaming the energy through space is no problem, and a microwave beam can penetrate even cloudy air easily. Like radar beams (many radars *are* microwave systems), microwaves can get through clouds, rain, fog, or snow without being blocked or scattered. Only a tiny fraction of the beam's energy would be absorbed by the atmosphere. Thus there would be minuscule effects on the atmosphere itself, and the beam would deliver virtually all its power very efficiently to the receiver.

The receivers themselves would look something like the television antennas that stick up from rooftops all around the world. But instead of single antennas, the receiving area would be a miles-wide "farm" of antennas. The term *receiving antenna* has been abbreviated into the new word, *rectenna,* and the area where the microwaves beamed from the SPS are received has been dubbed the *rectenna farm.* It could literally be a farm, according to Glaser and the SPS proponents. The rectennas would be spaced far enough apart so that the ground beneath them would receive sunshine and rain just like any other meadow or field. The microwave beam would be of such low intensity, spread out over a diameter of five miles or more, that it would pose no ecological threat to the crops being grown. When Glaser shows slides depicting the rectenna farm, there is always lush green grass growing between the metal poles, with cattle grazing on it.

Once received on the ground the microwave energy is converted back into electricity and fed into the power grid, just as if the electricity had been generated by ordinary ground-based power plants.

Each rectenna farm would receive five gigawatts of energy: that is, five thousand megawatts, or five billion watts, slightly less than one percent of the total US electric utility industry's installed capacity.

Five gigawatts is the equivalent of 210,000 barrels of oil *per day.* Ten Solar Power Satellites, then, could provide ten percent of the US's electrical power needs, without burning any fuel whatsoever and without requiring cooling water for ground-based power stations. Without polluting the atmosphere or water with waste heat, carbon dioxide or other combustion products, or radioactivity.

An SPS would be big. Huge. The five-gigawatt model would be the size of Manhattan Island: some 50 square miles of solar cells, power conversion equipment, antennas, and computers. But because the environment of space is free of wind and weather, and the SPS would be in zero gravity, the structure could be light in weight despite its enormous size. There would be no need to reinforce the structure in the way terrestrial buildings must be supported, and no need to provide vacuum chambers for the microwave generators and other components — they will be in a much better vacuum than anyone can create on Earth.

The first SPS would undoubtedly be constructed on Earth, transported piecemeal into orbit, and assembled in space.

The first steps in learning how to do this, working out the techniques and tools that will be necessary, can be done with the Space Shuttle. In fact, these techniques and tools are being explored right now — by the Russians, aboard their space station, Salyut 6.

According to Christopher C. Kraft Jr., director of NASA's Johnson Space Center in Texas (the famous Chris Kraft of Mission Control during the manned space flights of the 1960s), it will be necessary to develop a new booster, a Heavy Launch Vehicle, to lift the SPS components into geosynchronous orbit. The HLV would be an advanced form of Space Shuttle, totally reusable, with a first stage that can fly back and land at an airport, just as today's Shuttle Orbiter stage can. Capable of carrying payloads in excess of 400 tons, the HLV could haul a complete Solar Power Satellite into orbit in 200 flights. Figuring two flights per week, that is roughly two years; far less time than it takes to build a conventional power station on the ground — of one-fifth the power output.

The Heavy Launch Vehicle could be the major technical risk of the entire scheme. The SPS itself is based entirely on existing, known technology.

Tests have been made in the Mojave Desert and with large radio telescopes to measure how microwaves are transmitted through the atmosphere. The earlier calculations have been found to be accurate: The microwaves go through the atmosphere with very little interference or absorption, in practically any kind of weather.

Under contract to NASA, such prime aerospace companies as Grumman and Boeing have already developed key parts of the zero-gravity construction technology that will be needed to assemble or build Solar Power Satellites in orbit.

New spacesuits have been developed and will be tested in orbit by Shuttle astronauts. NASA will also test Astronaut Maneuvering Units in orbit — backpack units with cold-gas thrusters built into them that will allow the astronauts to move about freely during their extra-vehicular activity tasks.

The main technical unknowns concerning SPS stem mainly from the huge size and complexity of the beast. Its individual components are all based on existing, well-tested knowledge.

The costs of the first SPS would be just as large as everything else associated with it: Estimates have run from $20 to $60 billion. The larger figure includes development costs for the Heavy Launch Vehicle, which will be the workhorse for any and all space missions once it becomes operational.

Sixty billion dollars is obviously an enormous sum of money. But SPS backers, among them Jerry Grey, a research and engineering consultant to government and private organizations, insist that we will be spending that money on energy one way or another between now and the end of the century. The SPS may offer us the most useful return for the money. Each month the US spends nearly $10 billion on imported oil. In 1979, imported oil cost America $90 billion; in 1980, $100 billion. And the price keeps going up. These are the numbers that the SPS cost must be weighed against.

Engineering designs for the SPS are based on a nominal thirty-year lifetime for the satellite. Each SPS may last longer, but the engineers need a number on which to hang their cost estimates, and thirty years seemed reasonable, even if somewhat arbitrary. In thirty years, if American oil imports continue at their present level, and if the price of oil remains perfectly stable, the US will spend more than two trillion dollars (2×10^{12}) on foreign oil. One Solar Power Satellite delivering five gigawatts to the ground could produce income for its owners of more than $2.5 billion per year. This means that, over a nominal thirty-year lifespan, *the first SPS will amortize all the development costs of the system,* including the Heavy Launch Vehicle costs, and still produce a hefty profit for its backers.

It would take dozens of SPS's to replace the electricity now being generated by the some 270 million barrels of oil the US imports each month. Or fewer SPS's of larger power capacity. But even if the development costs run to $60 billion, SPS's backers insist that the risks are worth the eventual gain. In the long run, they claim, SPS will be far cheaper than any other energy source. Its up-front investment costs may be very high, but its long-term *operational* costs will be negligible. No fuel to buy, remember. And the second SPS will cost a fraction of the first. Finally, the environmental cost of SPS will be minimal — according to the Prometheans.

Not so! say the Luddites. "SPS is inappropriate technology," says Joe Foreman, an engineer in the Space System Division of the Naval Research Laboratory. "Not only is it a waste of money and resources and a threat to the environment, but also [it] damages the country's efforts towards energy independence by presenting a false solution to the country's energy problems." Foreman and others argue against each and every assertion made by SPS's backers, and take special issue with the claim of environmental "cleanness."

Microwaves may well replace radioactivity as the environmental bugaboo of the 1980s. Already the environmentalists are campaigning against the "zapping" of America with microwaves from television transmitters, telephone relay antennas, and microwave ovens. They have considerable cause for alarm, because prolonged exposure to a harmful level of microwave radiation can cause blood diseases and neurological disorders, and even can kill you.

There's an old story from the early days of the Distant Early Warning (DEW) Line of missile-defense radar stations that were put up along the Arctic Circle in the 1950s. Legend has it that Eskimos camped near the giant microwave antennas because they felt sheltered, somehow, from the cold when they were near the antennas. Then the Eskimos started dying, cooked by microwave radiation. The story is most likely a canard, but its moral is real: Microwave radiation can be harmful.

But how much is too much? What is a safe level of microwave radiation, a level that we can live with and not be harmed?

Which expert do you believe? In the United States, the government-accepted limit for occupational safety is ten milliwatts (ten thousandths of a watt) of microwave radiation over an eight-hour period. In Canada, it's one milliwatt, with a one-hour maximum. In the USSR and most Eastern European nations, the maximum is 0.01 milliwatt — a thousand times lower than the US and Western European standard.

Many environmentalists have pointed out that the Russians have done much more intensive work on low-level, long-term microwave effects than Western researchers have. And one wonders just why the Soviets have been beaming low-intensity microwave radiation at the US embassy offices in Moscow for so many years.

Thus, while SPS seems to offer clean energy, its environmental effects have not been fully explored. Not only would gigawatts of microwave energy be pumped through the atmosphere, but the beams could have deleterious effects on the upper atmosphere, messing up the ionosphere and ozone layers, some 20 miles and more above our heads.

Side effects from all this microwave radiation could cause problems on the ground, too, with radio transmission. The large sizes of the rectenna farms may be a drawback, and the high-voltage transmission lines needed to carry the power from the rectenna farms to the users could stir environmentalists' protests.

Tests have proved these fears to be exaggerated, if not groundless. Microwave radiation, in the amounts and intensities planned for SPS, will not harm the ozone layer of the ionosphere. And the microwave energy is absorbed by the rectennas — that's where the electricity comes from, after all — not by the atmosphere or the ground. At worst, the area of the rectenna farm will be slightly warmer than the region around it, a micro-climate effect similar to that caused by turning a farm into a suburban housing development.

There is plenty of empty and unused territory today that could be used as sites for rectenna farms. The first rectenna farm would most likely be placed in the American Southwest, perhaps on government-owned desert land, for example, the

White Sands Proving Grounds. That would provide an area with optimum weather conditions for the SPS's first full-scale tests, with no "right-of-way" problems over the real estate.

Eventually, rectenna farms would be placed as close as practicable to the end-users of the electricity, to minimize the length of the power transmission lines.

The exhaust gases belched out of the rocket engines used to boost the SPS components into orbit might have serious effects on the atmosphere. Studies have shown that the ozone layer could be literally punctured and riddled with miles-wide holes because of chemical reactions between the ozone and the hot rocket exhaust gases. On the other hand, Chris Kraft has pointed out that the 200 flights of the Heavy Launch Vehicle needed to orbit a single SPS would pump less carbon dioxide and water vapor into the atmosphere than seven months' worth of fossil-fueled power plants operating at the same five-gigawatt level. And while the rocket exhaust products would be limited to carbon dioxide, water vapor, and some oxides of nitrogen, the pollution products from fossil-fueled power plants contain tons of sulfur oxides and, in the case of coal, carcinogens and lung-damaging soot.

The favorite "pop" environmental disaster of bygone years was the Ice Age/Flood scenario, in which man's inhuman indifference to ecological morality resulted in either creating a new Ice Age or melting the polar caps and causing worldwide floods. But ever since it was learned that the chlorine and other compounds in spray-can propellant gases can deplete the ozone in the upper atmosphere, the Ozone Layer Threat has become the favorite worry of "pop" environmentalists.

I do not mean to denigrate the sober work of serious environmentalists; the worry is real. I have even switched from spray-can to roll-on underarm deodorant. But to those who fear that a hundred or so rocket launches per year will wipe out the ozone layer and allow solar ultraviolet to smite us all with cancer, I offer two points for contemplation:

1. The USSR launches a hundred or more boosters each year.
2. An end to cigarette smoking would avert more cancer than the weakening of the ozone layer will ever cause.

But there is an additional argument against SPS: its potential as a military weapon.

The microwave beam from a Solar Power Satellite will be kept diffuse enough so that it will pose no hazard to health, according to SPS planners. Of course, extremely prolonged exposure to even a very low-intensity microwave beam may eventually cause biological damage to animals. That is why the rectenna farms will be placed away from habitation centers.

But suppose an enemy power or a terrorist or a madman seized a Solar Power Satellite and swung its beam into the heart of a giant city? Or suppose the huge structure of the SPS were used to hide space weapons, such as nuclear bombs or very high-power lasers? Or (most likely of all) suppose the SPS's power were diverted from its peaceful application to power military death-ray–type lasers, either on the ground or in space?

The first possibility can be engineered out of the picture. It is a simple matter to build equipment into the SPS to simply shut it off if the beam drifts away from its rectenna farm. Even if the "bad guys" could get the beam going again, the people who would be depending on the SPS's power would notice very quickly that their electricity had been cut off — they would raise a ruckus long before the diffuse beam from the SPS could cause a dozen headaches among its targeted "victims."

The second possibility, hiding weapons aboard an SPS's huge structure, can be guarded against only by inspection. An international inspection system seems to be an inevitable by-product of SPS, just as modern jet air travel has led to international inspection of baggage for weaponry. It has even been suggested that this inspection function be taken entirely away from all political entities — including the United Nations — and handed over to insurance companies like Lloyd's of London. After all, they would be insuring the satellites and the personnel who work on them; it would be to the insurance companies' best interest to see to it that no weapons are brought aboard the satellites.

Using Solar Power Satellites to power other energy weapons, especially laser and other beam-type weapons in space, seems to me a very real possibility. But I doubt that SPS's built to deliver electricity to customers on the ground would suddenly be pre-empted by the military, except under the direst of circumstances. More likely, I suspect, the military will *itself* build SPS's for direct military applications. And by "the military," I do not mean merely American military.

The other side of the weaponry coin is vulnerability. Just when we're settling down to enjoy the electricity coming from SPS's, this scenario runs, somebody either shuts them down or destroys them. I even wrote a novel about this, back in 1978, titled *Colony.*

Solar Power Satellites may well be vulnerable to what the think-tank futurists now call SMAT (sabotage, mutiny, attack, and terrorism). While its enormous size would make an SPS difficult to destroy totally, it would be comparatively easy to knock out such key components of a Solar Power Satellite as the microwave transmitter. Or the crew. Again, international inspection and patrol seems the only answer, short of out-and-out military guards.

In their hearts, though, most Luddites feel these scenarios are strictly science fiction. They insist that SPS is much too expensive ever to get off the ground. Nobody's crazy enough to spend $20 to $60 billion on giant energy satellites, they maintain. In fact, they see the entire SPS idea as a giant swindle — a way to pour titanic sums of taxpayers' money into the coffers of the big, bad corporations.

The centralized-vs.-decentralized battle arises here. The Luddites recognize that it will take an enormous concentration of government money and corporate know-how to build Solar Power Satellites. They want that money spent on ground-based, decentralized solar energy systems, such as active and passive solar heating equipment and solarvoltaics. The Luddites insist that such systems are better for all, particularly since they are decentralized and give the individual a chance to get out from under the controls of governments and corporations. They fear that if the SPS program begins to receive the billions of dollars that it needs, the money will be siphoned away from the ground-based solar energy systems that they want.

If the Luddites see the Establishment's Big Government and Big Business as the enemy, then the SPS is the enemy's sword, pointing straight at their throats. They have no hope for the kind of government and corporate support they desire if SPS moves from paper studies to hardware engineering.

On the other hand, among some Prometheans, the Solar Power Satellite is admired not merely for its own charms, but because of its "connections." As we have seen, to build the first SPS will take a huge investment in developing the Heavy Launch Vehicle, tons of space hardware, large teams of highly skilled and precisely trained men and women, ranging from ground-based engineers to construction crews that will live and work in space for months at a time. All of these investment costs can be applied to other projects, once they have done their job for SPS. In fact, it makes no economic sense whatever unless they are applied to other jobs in space. We cannot afford to repeat the mistake of Apollo and disband the thousands of dedicated people and scrap the intricate hardware that we have labored so hard to build. The Prometheans, then, see SPS not only as a goal in itself, but as a means to establish a strong and permanent capability to build large structures in space.

There is an old Arab tale about a Bedouin who is awakened from his dreams one night to see his camel's nose sticking into his tent. "It's only the camel's nose," the Bedouin thinks, and goes back to sleep. Before long the entire camel is sleeping in the tent alongside its master and there is no way the Bedouin can get it out.

Both the Prometheans and the Luddites see SPS as the nose of the camel. The Luddites fear that this first step toward truly large-scale operations in space will crush their dream of a decentralized Solar Democracy. The Prometheans hope that SPS will open the way to a truly powerful space program. Neither side has been able to answer the basic questions about SPS: Will it work efficiently? How much will it cost? How soon can we put one into operation?

After Glaser's first published paper on the SPS idea, in 1968, the concept slowly gained credence within the space community. Slowly, quietly, paperwork studies of various aspects of SPS were started at NASA. The late 1960s and early '70s were not a time of daring and high budgets in NASA. Apollo was being garroted; new programs were being aborted. Still, by 1974 there was panic even in Washington over the energy crisis, and the SPS concept was one way NASA could show it was being responsive to public needs.

By 1977 the Office of Management and Budget transferred responsibility and funding for SPS from NASA to the Energy Research and Development Administration, which was soon to be reshuffled into the newly created Department of Energy. The Office of Management and Budget, often called by its acronym, TOMB, is the White House's purse string guardian. Its staff consists mainly of accountants, usually called "bean counters" by people in operating departments of the government. TOMB's job is to get the most out of the taxpayers' money. Too often this means being penny-wise and pound-foolish.

It often looks good, and therefore is politically tempting, to cut the funds for a program that has no immediate, practical payoff. We have already seen how a lack of foresight shut down magnetohydrodynamics development in the US. Unless there is a strong leader in the Oval Office who understands the necessity for longer-range goals, it is all too easy for TOMB to kill programs outright or to starve them to death. We will see more of this problem later, in the chapter discussing the Space Shuttle.

The Solar Power Satellite became something of a political football, kicked between DOE and NASA. The neat solution that the White House bean counters had decided on was to give the funding to DOE and have them contract with NASA for the necessary studies. The reasoning was: DOE is responsible for energy developments, so SPS should be under their aegis. NASA has the technical expertise for SPS, so Energy will go to them for the work.

Like the neatly typed strategies of many a football coach after a game against the Pittsburgh Steelers, the White House's plan for SPS got mauled pretty badly.

There is a syndrome throughout government and industry called "NIH": Not Invented Here. The Solar Power Satellite was a NASA idea, and DOE's people had scant understanding of such a "far out" concept. Nor much enthusiasm for it.

When SPS was transferred to Energy, that department's big scheme for solving our energy problem was to pump a year's worth of oil *back into the ground* to provide a strategic reserve in case of another Arab oil embargo. The plan was to cost $20 billion, just about as much money as it took for the whole Apollo program. And the plan collapsed when the Saudis raised a collective eyebrow and let it be known that they didn't like the US having a year's worth of strategic reserve.

Twenty billion dollars, to provide twelve months' worth of oil reserves. If that money had been invested in high technology, we could be well on our way to a prototype SPS in orbit. *Plus* a healthy energy technology program that embraced ground-based solar, MHD, and other options.

But DOE's idea of high technology is coal gasification and synthetic fuels. SPS was kept in the paper-studies category, receiving a few million dollars per year and constantly subjected to wrangling between the budget committees of the House of Representatives and the Senate.

Glaser, meanwhile, had not been altogether idle. In April 1978, the formation of a nonprofit, privately funded Sunsat Energy Council was announced, with Peter Glaser as president. Sunsat's members included such aerospace companies as Boeing, McDonnell Douglas, and Lockheed; such electrical power equipment manufacturers as General Electric and Westinghouse; and utilities like Southern California Edison. Sunsat's legal counsel is former Senator Frank Moss, and at the press conference announcing formation of the nonprofit Council, Senator Henry Jackson (D., Wash.) warmly endorsed the idea.

But the Washington arena is not unanimously in favor of SPS, nor of Sunsat. Representative Richard Ottinger (D., NY) said, "It's not bad enough the space-industrial complex is trying to slip one over on the taxpayer and solar energy enthusiasts, but now they're trying to put one past the IRS."

The battle seethes to and fro. In President Carter's Fiscal Year 1981 budget, SPS studies were initially funded at about $5 million. But in the frenzy of budget cutting that took place just before the Congress adjourned to go politicking in midyear, *all* SPS funds were sliced out of the newly "balanced" budget by TOMB and the White House. Prometheans around the nation reacted with a roar, and the new Congress that convened in January 1981 was faced with conflicting, often contradictory, and increasingly emotional attacks from both camps.

SPS has become a *cause célèbre,* and in the heat of the argument over it, the basic question of whether or not it is an effective way to solve our energy problems is being overlooked.

Increasingly the Luddite argument concentrates on the dangers of SPS. Visions of Three Mile Island and Love Canal do a *danse macabre* in the Luddites' heads when they think about that microwave beam.

That is where the SPS concept stands now, in the middle of the struggle between a decentralized Solar Democracy and a highly centralized Corporate America. If the Luddites have their way, the concept will be studied to death and quietly forgotten. If Prometheans win, we could see at least one SPS glittering in the sky before the end of the century.

Somehow I am reminded of the decision made by the US Army Quartermaster Corps, more than a century ago, regarding repeating rifles. This new kind of weapon could fire shot after shot, without reloading. The Quartermaster Corps studied the new type of rifle carefully and decided not to buy it. Why?

Economy. Repeating rifles would tempt the soldiers to become spendthrift with their ammunition. If they have to reload after every shot, the bean counters reasoned, they will make each shot count, thereby saving the taxpayers' money.

So the Seventh Cavalry went to Little Big Horn with good, reliable, single-shot rifles. The taxpayers saved a lot of money. But Sitting Bull's braves, who had no chance to study the problem as carefully as Washington, had purchased repeating rifles. *Sic transit* Custer's Last Stand.

Ridiculous? Then how about the study done by the Department of Defense in the late 1960s proving that very high-power lasers (which were then in the laboratory development stage) would not replace machine guns as infantry weapons? I was in on that one. Someday a squad of American infantrymen are going to have the machine guns melted out of their hands — and their hands along with them — by high-power laser weapons.

It was Norman Cousins who said:

The world will end neither with a bang nor a whimper, but with the strident cries of little men devoted to cost-benefit ratios. If cost-benefit ratios had governed our history, Socrates would have become a baby-sitter, Newton an apple polisher, Galileo and Giordano Bruno court jesters, Columbus would have taken out a gondola concession in Venice, Thomas Jefferson would have become a tax collector, John Milton would have written limericks, and Albert Einstein would have changed his name and stayed in Germany.

21

Workshops in the Sky

> We cannot escape from the past, but neither
> can we avoid inventing the future.
> —RENÉ DUBOS

In the wide and starry band of near-earth space, beginning about
200 miles up and extending to 22,300 miles, where a satellite can be
placed in stationary orbit rotating in unison with the earth . . . [there
is] the possibility of an industrial bonanza. Operating in this pure
and virtually gravity-free environment, factories could produce
novel materials worth as much as $30,000 a pound back here on
earth.

A brochure put out by NASA? Press puffery for a new science
fiction movie? No, it's neither. The quote above is from *Fortune*
magazine's January 29, 1979, issue, the first installment of a
two-part feature by writer Gene Bylinsky. "No corporation
affected by changes in technology can afford to ignore the new
era of innovation that is about to begin," Bylinsky concludes,
after citing the dozens of new industries that can be built in
space.

Over the past several years I have discussed the possibilities
of industrial operation in space with a wide variety of audiences, ranging from high school science students to executives

of major corporations like IBM, U.S. Steel, RCA, Alcoa, Rockwell International, and others. The students are naturally afire to reach into space and carve out careers for themselves in what they perceive to be an exciting new frontier. What surprised me most was that many corporate executives are equally enthusiastic about space industries. Few of them hope to go into space personally, but they are quick to grasp the economic potential of space industries. They recognize the risks, certainly, but they also perceive the potential profits.

There are always a few in every audience, regardless of age, who simply cannot accept the idea of working and living off the Earth, of building factories in orbit and making profits from space manufacturing. Wild science fiction ideas, they say. Sure, maybe in a hundred years or so somebody will build a space station in orbit. But that's 'way down the road. We've got enough to worry about right now without getting worked up over crazy space-cadet stuff.

First I tell such scoffers that there is already a space station in orbit, and there are two cosmonauts working in it right now. Then I refer them to the Fortune 500, the list of the top 500 manufacturing firms in the US, compiled each year by the editors of *Fortune* magazine. Twenty-five years ago, when visionaries like Arthur C. Clarke, S. Fred Singer, Wernher von Braun, Frederick C. Durant III, and others were urging government and industry to move into space, most of the industrial executives of the giant corporations scoffed at the idea.

Compare the position of those scoffers in the Fortune 500 list of 1955 with the list of 1980. Many of their companies no longer exist. Whole industries, especially steel, rubber, and textiles, which were the mainstays of American heavy industry in 1955, have slid down the list steadily, becoming less profitable and losing more and more of their business to foreign competitors. Not because they scoffed at space missions, but because they *did not keep up technologically.* The American automobile industry is teetering on the brink of the same pit today, desperately trying to catch up with the technological innovations of the foreign auto manufacturers to recapture the American market.

"The high-technology companies have made some of the most conspicuous gains in sales over the past twenty-five years," says Linda Snyder Hayes, in her review of the Fortune 500's first quarter-century. Xerox, Hewlett-Packard, and Polaroid were not on the original 1955 list. Control Data, Digital Equipment, Memorex, Data General, and Storage Technology — all computer companies — did not even exist in 1955.

Cosmetics and pharmaceutical companies, heavily dependent on chemical technology, also "staged impressive performances over the past twenty-five years."

The champion money losers of the quarter-century have been Anaconda Copper, Singer Corporation, Bethlehem Steel, and Chrysler — which lost a whopping $1.7 billion in 1980, and $1.1 billion the year before.

The lesson of the Fortune 500 is clear: Companies that ignored high technology suffered losses, went broke, or were absorbed by their more profitable competitors. Companies that capitalized on high technology — such as the computer and electronics firms — prospered and grew.

I am not saying that a steel firm like Jones & Laughlin was bought out by LTV Corporation (an aerospace company) because of the space program. As a matter of fact, aerospace firms barely held their own through the past twenty-five years, mainly because most of them have virtually only one customer — the US government — and rise or fall on the strength of federal spending in aircraft and space.

Rather, an aerospace company could buy out a steel company because high technology produces profits. Around the industrial world, high technology is the key to corporate success. Since aerospace technology is at the very cutting edge of knowledge in such vital new areas as electronics, materials processing, systems organization, fuels and energy research, aerospace technology is the spearhead of new industries. Companies that ignore the new potentials of space industries are turning their backs on the developments that will determine which firms are listed in the Fortune 500 of A.D. 2000.

At first thought, the idea of building factories in space, and making profits from it, seems outlandish. But every new territory seems strange, hostile, and most likely useless. More than a century ago, when the US Senate was debating the question of appropriating funds to build a railroad across the continent, the doughty Daniel Webster proclaimed:

> Mr. President, what do we want with this vast worthless area? ... What can we ever hope to do with a western coast of 3000 miles, rock bound, cheerless, uninviting and not a harbor in it? ... I will never vote one cent from the public treasury to place the Pacific coast one inch nearer Boston than it now is.

Even earlier, in 1805, Zebulon Pike, the intrepid soldier and explorer for whom Pike's Peak is named, led a small expedition across the Mississippi and saw the Great Plains for the first time. "The Great American Desert," he called it, firmly believing that the vast treeless plains would never have any human use. Today the Great American Desert is the wheat and corn bowl for the world.

The environment of space is strange and decidedly hostile to human habitation. But it is far from useless. When an industrialist looks at the space environment objectively, he sees great advantages there. As G. Harry Stine pointed out in his book, *The Third Industrial Revolution* (Ace Books, 1979), the space environment offers four advantages that are extremely attractive for industrial operations: virtually free energy, controllable temperature extremes, very high vacuum, controllable gravity.

Energy and Temperature: An increasingly stringent problem on Earth, energy is abundant in space, almost embarrassingly so. The Sun shines constantly, and solar energy can be converted into electricity in any of several ways or used directly as heat. Ever since the first Vanguard satellite was orbited, photovoltaic cells have routinely converted sunlight into electricity for spacecraft.

Not only is electricity easy to produce, but the Sun's heat can be used for a wide range of industrial processes. With simple mirrors it is possible to focus sunlight and attain temperatures of thousands of degrees. All the smelting, metalworking, chemical processing, boiling, heating that is done at an orbital factory can be done with direct or concentrated sunlight. Much of this processing can take place out in the open, rather than in an enclosed chamber. Thanks to the effective weightlessness of orbit, and the vacuum of the space environment, technicians can "hang" an ingot of iron, for example, outside the space factory, focus a parabolic mirror on it, and melt it down quite easily. No drips, either. The molten ingot will hang there just as weightlessly as it did when it was solid, and slowly assume a spherical shape, due to internal tension forces.

Not only can a space factory easily attain very high temperatures, but very low temperatures as well, simply by shielding a volume of space from sunlight. A well-shadowed region could be cooled down to the "ambient" temperature of empty space, close to Absolute Zero. Because vacuum is an excellent thermal insulator (the secret of the Thermos bottle) a space factory could be melting steel ingots in one place and, only a few yards away, could simultaneously be liquefying hydrogen or helium.

In sunshine it's hot; in shadow it's cold. Temperatures can be manipulated up and down the scale, from thousands of degrees to almost Absolute Zero, merely by arranging the amount of sunlight or shade: without burning a gram of fuel; without building heaters or refrigerators; without separating the hot work from the cold work by more than a few yards. Virtually free energy means freedom from a heavy and continuous expense, as well as freedom from the pollution products that inevitably accompany energy systems on Earth.

Vacuum and Gravity: For much of my adult life I worked in the aerospace R&D industry, and the sound that still haunts my dreams is the clatter of vacuum pumps. It costs a lot of money and time to make nothing — on Earth. So many industrial processes require vacuum chambers at some stage of their operation that a considerable part of the cost of electronics, pharmaceuticals, metals, and other industrial products stems from the need to pump air out of a chamber and produce nothingness.

Space is a very good vacuum. Just a few hundred miles above our heads is a better vacuum *for free* than any we can buy at any price on Earth. Combine this excellent vacuum with zero effective gravity and you have the possibility of "containerless" processing, which can lead to the routine manufacture of ultrapure materials.

On Earth, when you want to mix various ingredients you must put them into a container: a bowl, a centrifuge, a test tube, a blast furnace. No matter how well you stir the ingredients, on Earth, the heavier ones will always tend to sink to the bottom. And there will always be microscopic bits of the container mixed in, too. While for most processes this contamination does not matter, in certain industries, such as pharmaceuticals, purity is an important quality.

In space everything changes, for the better. A space factory does not need containers: The materials to be worked with can hang in vacuum, weightless, unsullied by impurities. And since all the ingredients in any mixture would be equally weightless, there would be no "heavier" elements to sink to the bottom. Indeed, there is no bottom! All directions are the same under zero gravity. We can sit on the ceiling or the walls just as easily as on the floor, or hover in the middle of the room, if we choose.

Science fiction writer Alfred Bester put his fertile imagination to work on the advantages (and problems) of cooking in space. *Voici:*

> The preparation of the beef filets is perhaps the most typical aspect of cooking in free-fall. The chef removes them from the fridge and poises them an inch before the broil plate of the range. There they float while he adjusts the heat of the sun to sear them . . . No pots, pans, gridirons. Everything can be floated before the stove, even liquids.

These four advantages of the space environment add up to a very favorable climate for industrial operations in orbit — if you can get into orbit cheaply enough. This underscores the crucial importance of NASA's Space Shuttle, which is aimed at reducing the price of going into space.

The Russians have already tested several space manufacturing processes aboard their Salyut 6 space station. In the US, the only such tests we have carried out were conducted aboard the Skylab in 1973–74, and in conjunction with Russian cosmonauts during the very brief Apollo-Soyuz joint mission in 1975. We have done paper studies, of course. We have tons of paperwork. For what it's worth, the paperwork shows such promise from space manufacturing that many industrial corporations are ready to move ahead with tests in orbit, as soon as the Shuttle can loft their equipment into space.

As a former marketing executive for the research division of a Fortune 500 corporation, I ask myself: Just what areas offer the best business opportunities in space? A second question immediately follows: How much money can we make Up There?

The most immediate opportunity in space is in the area of communications. Already, companies like Comsat and Intelsat are doing about a billion dollars' worth of business per year, through operating communications satellites. RCA, Western Union, AT&T are all in the business. Telephone and television messages are now routinely relayed across continents, across oceans, across the whole globe by satellites.

Such communications are only one part of a broader market area called *information services*. In studies done for NASA, several private analysts have found many information services that can be done easily and profitably from space, but only with great difficulty (or not at all) from the ground. They include:

Wristwatch-sized communicators that combine the functions of a computer, telephone, instant reference service, library, navigational locator, and (oh, yes) even a calendar/watch. Such a device could be built today, but the tremendous message load from hundreds of millions of these gadgets will require very sophisticated satellites to perform as switchboards and relay stations.

Today's communications satellites are relatively simple relay stations; the complex electronics remain on the ground, at the sending and receiving stations. But the microchip revolution is starting to produce hardware that is microscopically small, and this is leading to a new concept in comsats. Tomorrow's comsats will be very sophisticated "smart" switchboards in the sky, handling thousands of messages simultaneously. The individual sending/receiving communicators that you and I wear on our wrists will be simple, rugged, and cheap. The complexities, and the expense, will go into the automated satellite relay stations. There will be millions of wrist communicators on the ground; only a few "smart" comsats will be needed to handle their calls. This kind of market could grow to $20 billion a year, or even more, over the coming twenty years.

Mail service could become electronic, highly automated, and relayed by satellite for same-day — or even same-hour — delivery. The Satellite Business System is already working for private companies. In the relatively near future, all mail could be delivered electronically. Those who worry about privacy can use the old hand-carry system; but for speed, electronics is the answer.

Voting could also be carried out electronically, which may lead to direct polling of the electorate on sensitive issues. Instead of waiting until your representative comes up for reelection, you can instruct him or her, vote by vote, issue by issue, thanks to electronic polling.

Information services also include the kinds of observational satellites that are now in orbit — for example, Landsat — as well as weather satellites, navigational satellites, and satellites that could monitor air traffic, railroad systems, even highway conditions all across the nation. In all, the information services market promises to yield some forty to fifty billion dollars' worth of new jobs and profits *per year* within the next twenty to thirty years.

Energy is another market area that offers tremendous potential for space operations. We have already discussed the possibility of building Solar Power Satellites and of disposing of nuclear wastes in space. It may also be possible to build large, lightweight mirrors in space to redirect sunlight to areas that would normally be dark. The growing season in high-latitude lands like Alaska could be extended for weeks or perhaps even months in this manner. The length of day could be extended anywhere that farmers need more growing time for their crops. This could lead to a dramatic increase in crop yields for marginal farming areas in Asia and Africa. It could also lead to gross ecological unbalancing, though, and very detailed studies will have to be undertaken before such "redirected insolation" experiments are attempted in the real world. But the technology is available, and it would be relatively inexpensive.

The truly gung-ho pronuclear enthusiasts insist that we must go ahead and develop nuclear breeder reactors. Breeders are reactors that use low-grade fuel like unenriched uranium and thorium and produce not only energy, but enriched uranium and plutonium as a by-product. If we must go full-scale into nuclear fission power, the breeder is a way of stretching the world's supplies of uranium and other fissionable elements almost *ad infinitum*. But breeder reactors are feared even more than the ordinary reactors used in today's nuclear power stations, because breeders produce more radioactive material, including the deadly and politically volatile plutonium. Plutonium is the stuff of cheap nuclear bombs, and Prometheans and Luddites alike would sleep more soundly at nights if they knew there was no plutonium around to be used by some unstable dictatorship or terrorist group.

If the energy situation becomes so bad that we are forced to develop breeder reactors, why not build and operate them in space? Their main function would be to produce fuels for existing nuclear power plants on the ground. The energy they produce could be beamed to the ground just as the energy from Solar Power Satellites would be.

An orbital "quarantine" for breeder reactors could make our environment here on Earth much safer from a nuclear accident. And it could make it much more difficult for spies or saboteurs to hijack the plutonium produced by the breeder or to demolish the breeder itself.

In Chapter 20 we spoke about the very high initial costs of Solar Power Satellites. But when we look at the market for energy products from space, those costs become income for somebody — lots of somebodies, the firms and the people who build and operate the SPSs. If SPS is pushed as hard as it could be, we could see a market of up to $100 billion per year in energy-related space operations by the year 2000, according to economic analyses done by Science Applications, Inc., and NASA. An additional billion dollars per year could be spent on nuclear waste disposal, and a decision to build breeder reactors in orbit could add another $10 billion per year to the energy market.

Fantastic numbers? Not if you realize the scale of today's energy market. The US is now spending about $100 billion per year to import foreign oil. Exxon's sales in 1979 were $79.1 billion, while General Motors (in a poor year) racked up $66.3 billion in sales. The numbers are not fantastic; the size and power of our economy is. We can put that power to work for us in space.

In addition to information services and energy, there is a vast potential market in actual space manufacturing: producing finished products for use (and sale) on Earth. Orbital factories could have great advantages over ground-based manufacturing facilities. Under the weightless, airless conditions of orbit, it is possible to grow crystals more uniformly and many times larger than they can be grown on Earth. Organic chemicals and biologicals can be separated and sorted out more easily in space. New types of glass, metal alloys, and other materials can be created in space factories. Studies have shown that new kinds of steel, ten to 100 times stronger than any made on Earth, can be manufactured in orbit. TRW Corporation has identified 400 new alloys that cannot be made on Earth at all, but can be produced in space. Superconducting magnets, self-lubricating alloys, ultrapure optical lenses and glass fibers — the list of space-manufactured products is long and growing. Gallium arsenide, a critical material for lasers, microwave transmitters, and light-emitting diodes, sells for $15,000 a pound today. It can be made more cheaply and easily in a space factory. So can the plastic microspheres needed for laser-fusion experiments, which now cost $30,000 per pound.

As we will see in later chapters, the raw materials for these factories may well come from the Moon and other bodies of the Solar System, rather than from Earth. It is very costly in terms of energy, and therefore money, to lift payloads from Earth's surface. But it is quite easy to send payloads of finished manufactured goods back to Earth. Thanks to our planet's thick mantle of air, a re-entry vehicle can coast down from orbit with the barest of pushes and glide through the atmosphere to a safe landing wherever the cargo manifest calls for. A landing in Peking would cost no more than a landing in Vienna, or anywhere else. Every place on Earth is only a few hundred miles away from an orbital factory.

Industries as diverse as pharmaceuticals, electronics, optics, metals, even jewelry can all benefit from the advantages of near-zero gravity, high vacuum, and precise temperature control to be found in orbit.

And if orbital "quarantine" is the best place to tinker with nuclear breeder reactors, it may also be the safest location for biological laboratories involved in recombinant DNA experiments and genetic engineering. If the molecular biologists inadvertently produce a dangerous mutation while "manufacturing" microbes for industrial uses, there is no chance of the mutated microbes contaminating the environment of Earth.

Conservative estimates of the market potential for space manufacturing have ranged from $5 billion per year on upward, over the coming twenty years. Clearly this is an area that is limited only by the imagination — and courage — of our industrial leaders.

Space-based manufacturing may even be of inestimable help in producing the Solar Democracy that the Luddites cherish. Studies have shown that orbital manufacturing techniques can reduce the cost of solarvoltaic cells to about one percent of their cost on Earth. Thus the chances are that if we ever achieve a decentralized Solar Democracy, with every home generating its own electricity from rooftop solar cells, those cells will have been manufactured in orbital factories.

All of these activities in orbit will require a great number of men and women shuttling back and forth. Construction workers, scientists, physicians and nurses, engineers, technicians — hundreds, if not thousands of people will go into orbit for periods of a few weeks to a few months. They will be in space to work, just as an oil rigger might go to Alaska's North Slope or an offshore platform in the North Sea. Inevitably, as these workers come back to Earth and show that they have lived in space for months with no ill effects, tourists will want to get into orbit. Today's jet set will become tomorrow's rocket set.

The economics look favorable. In a typical year of the 1970s, between six and seven million Americans spent vacations overseas. More than 70,000 of them paid nearly $8 million per year for luxury cruises aboard ships such as the *Queen Elizabeth II.* (Even Isaac Asimov, who absolutely refuses to set foot in an airplane, crossed the Atlantic on *QE II.* Twice.) A round-the-world cruise aboard *QE II* costs roughly $25 per pound of passenger weight. If the cost of shuttling into orbit can be reduced to $25 per payload pound — as many aerospace engineers believe it can be, by the 1990s — then we might assume that tourist facilities in orbit would get roughly the same number of passengers as today's round-the-world cruises handle. Perhaps even more.

If 50,000 tourists per year visit an "Orbital Hilton," we could estimate an annual market of more than $100 million. Not all that much, compared to the billions to be made in manufacturing, energy, and communications services. But certainly nothing to sneeze at. Fifty thousand tourists per year means 962 tourists per week, or 137 per day. A tidy little business. Remember the scenes of the Stanley Kubrick–Arthur C. Clarke film, *2001: A Space Odyssey*, where a very sleek shuttle rocketplane bearing PanAm markings flew to the space station? Airlines may well become spacelines by then (several are already considering it), and Hilton will not be the only hotel chain vying for space tourists' money.

The potential revenues from all these space industries — communications, energy, manufacturing, tourism — are truly staggering. Even though it is rather too early to attach solid dollar figures to this potential, the order of magnitude is clearly in the tens or even hundreds of billions of dollars. Per year. We are not speaking of government tax money now. This is not a make-work, pump-priming, tax-draining program out of Washington. The hundreds of billions to be made in space will come from private capital investment, which will quickly be turned into profits for the investors.

Certainly the government will play a role in this developing market, just as the government played a role in developing the railroads, the airlines, the communication industry, and Comsat. But no government can or would develop the markets we have examined with the speed or determination of private, profit-driven companies. If we want to develop the space market, if we want to add a hundred billion dollars per year to our gross national product by the year 2000, if we want to produce millions of new jobs — jobs that do not exist today — then we must turn to the private investors, the profit-seeking corporations.

Many Americans are skeptical of private enterprise. Not only the Luddites fear the power of Big Business to ride roughshod over the individual citizen. In the minds of many, the words *corporation* and *ripoff* are inextricably linked. But there seems to be no way to avoid centralization in an undertaking as large and complex as the new space program. To do a big job we need powerful tools. To develop the incredible potential of space *quickly enough to solve the problems of Earth* we need the fast-moving corporations, large and small, that have access to the resources of personnel and funding that are needed.

The corporations are the key. They are the driving force in the American economy, like it or not. Without their investment there can be no new space program nor, indeed, any effective program of any kind in American society. (The American automobile has been widely criticized. But what would the auto industry be like if the Department of Transportation was the only manufacturer of automobiles in the US?)

The first phase of the space program, the explorations that culminated with Apollo and the unmanned probes of Mars, Saturn, and the other planets, was directed by a government agency: NASA. Industrial contractors worked under NASA's guidance. In this new phase of the space program, where we want to use the advantages of space to make jobs and profits on Earth, it will be necessary for private industry to risk its own investment capital, just as private industry has done in the electronics, pharmaceutical, aircraft, and so many other commercial markets.

The alternative is to keep the space program strictly a government operation. This is exactly what the Soviet Union and other socialist nations are trying to achieve. They have produced a treaty in the United Nations that would effectively bar private enterprise from space. If they succeed in this, if we allow them to succeed, we will probably never see the benefits that can come from space.

We will consider the UN's Moon Treaty in a later chapter. Its significance to us at the moment is that the profits from space industries are real enough so that the rest of the world is attempting to create laws that will govern how those profits will be split up among the nations of the world.

Meanwhile, some private entrepreneurs have already tried to start their own space programs. A West German company, OTRAG, actually leased a huge tract of land in equatorial Zaïre and began launching test rockets. The aim of the company (the name translated into English means Orbital Transport and

Rockets, Inc.) was to launch commercial payloads into orbit for profit, just as a commercial airline company carries passengers and freight around the world.

Maybe it was because OTRAG is a German company, or maybe it was because they simply do not want private enterprise in space: Either way, the socialist nations of East Europe and Africa attacked OTRAG with an intense propaganda offensive and, finally, with armed force. The Soviet propaganda pictured OTRAG as a thinly disguised Nazi military operation, testing nuclear missiles in Africa's heartland. In 1978, Katangan rebels invaded Zaïre from neighboring Angola, where they had been armed and trained by East German "advisors." Aided by Belgian and French Foreign Legion troops, Zaïre beat back the invaders, but it was clear that the German rocket base was a source of intense displeasure among other African nations. OTRAG was expelled from Zaïre; its lease on the launching area was canceled.

In late 1980 OTRAG resurfaced with an agreement for a launching base in southern Libya. Whether this will work or not remains to be seen.

American private entrepreneurs are seeking to raise capital for their own rocket-launching businesses. Like OTRAG, they want a launching base near the Equator, because the Earth's greater spin at the Equator means a rocket can reach orbit on less thrust, which makes operations that much less expensive. Interest among potential investors is high, but no one has yet reached the stage of leasing territory or building rockets.

What is certain is this: Private enterprise must be involved in the new era of space development, if the benefits from space are to be produced in our own generation, in time to forestall the collapse envisioned for the early twenty-first century. And to accomplish this, the most important *technical* step is to reduce the cost of reaching orbit — which brings us to the critical importance of the Space Shuttle.

22

Shuttle: The Egg Basket

> Even the longest journey is begun with a
> single step.
>
> —CONFUCIUS

To get into space, to build the Solar Power Satellites and space factories, to mine the Moon and asteroids and create the space industries that can make everyone on Earth richer than maharajahs, we must first get off the ground. Take the first step on the long journey.

The first step into space is the most difficult. Lifting a payload from Earth's surface into even a very low satellite orbit takes a huge amount of energy, which means it costs a huge amount of money.

The first step in our new space program is the Space Shuttle. But like the first step of a baby, the Space Shuttle effort has been hesitant, weak, and marked by delays and failures. Like a baby being urged on by too many overeager adults, the Shuttle has suffered from confusion of purpose and conflicting pressures.

The original idea behind the Shuttle was solid common sense. Space flight costs too much to be practical for industrial or commercial purposes. Only programs backed by a very powerful government, such as military programs, or efforts carrying the image of national prestige, such as Apollo, can afford to go into space on throwaway rocket boosters.

Physicist Theodore Taylor calculated the cost of ordinary jet airliner service between New York and Los Angeles under the ground rules that: (1) there is only one flight per month; (2) the plane is thrown away after each flight and a new one is built; and (3) the entire cost of Kennedy and Los Angeles airports is included in the ticket price. Under those rules, the cost of traveling between New York and Los Angeles is approximately the same as the cost of putting payloads into orbit.

Even under these commercially unattractive conditions, communications companies like RCA are quite willing to pay NASA $50 million to launch a relay satellite, because the comsat saves RCA hundreds of millions of dollars in ground-based relay towers or transoceanic cables. But if U.S. Steel, for example, wanted to build a steel mill in orbit, the cost of boosting up the personnel and materials would be backbreaking — far too high to make the scheme practical. *If* the job was to be done by one-use, throwaway rocket boosters.

Hence the Shuttle, a *reusable* transportation system that takes off like a rocket, carries some 30 tons of payload into orbit, and then returns for another mission. And another. And another.

The Space Shuttle consists of two main components: the booster and the orbiter. Originally, the booster was to have been a returnable vehicle that could fly back to Earth under its own power and land like an airplane. But budget cuts and program stretchouts forced NASA to turn the booster into an only partially recoverable set of rocket engines and a fuel tank. The orbiter, mounted piggyback on the booster, looks something like a flat-bottomed, delta-winged airplane, about the size of a DC-9 airliner. It has three liquid rocket engines for propulsion, plus maneuvering rockets for use once it is in orbit.

The orbiter is about 124 feet long. Its wingspan is a stubby 80 feet. Its cargo bay, where the payloads are carried, is 15 feet wide by 60 feet long: almost as long as a tennis court and a bit more than half as wide.

The booster consists of two large solid-propellant rocket engines and, carried between them like a huge egg, the big external tank that holds liquid hydrogen and liquid oxygen for the orbiter's three rocket engines. The solid rockets develop 1760 tons of thrust apiece; each of the orbiter's three liquid rocket engines produces 235 tons of thrust. Altogether, the Shuttle lifts off with the thrust of 4225 tons.

The solid rockets are dropped off in flight once their propellants have been consumed. Parachutes lower them to the ocean, where recovery teams bring them back for refurbishment and future use. NASA engineers estimate that the solid booster engines can be used about ten times each.

The external propellant tank (which is considerably bigger than the orbiter itself) stays with the orbiter until they reach orbit. Then it is either fired back to re-enter the atmosphere and burn up, or it is released gently to remain in orbit for possible later use as building material.

The orbiter can remain in space for as long as 30 days, although most of its early missions will be only a week or so in duration. During re-entry, the orbiter can maneuver about 1000 miles across country, east–west. That means the crew can select a wide range of landing fields.

The orbiter's heat shield must be refurbished after each mission. The thousands of thermal tiles used on the first Shuttles were the source of many a nightmare among the technicians and astronauts, because of the difficulties in making them stick to the orbiter's skin. Other heat shield systems are being developed for later Shuttles.

NASA estimates that each orbiter will be good for 50 to 100 missions before it must be scrapped. Plans are already under way, however, for 500-mission Shuttles.

The Shuttle crew consists of two pilots, a systems monitor (similar to the flight engineer on an airliner), and a payload specialist, whose duty is to get the payload into orbit properly.

In essence, the Shuttle is like a durable pickup truck, built to carry sizable payloads into low earth orbit (LEO) for less than half the cost of using throwaway boosters like Saturn, Titan, Delta, or other expendable, hand-fashioned "sportscars."

That first step into space, the first hundred or so miles between Earth's surface and LEO, is the most difficult part of any space journey. Earth is a rather massive planet, with a strong gravitational field holding everything pinned down to its surface. You would have to travel to the giant planet Jupiter before you find a gravitational field as strong as Earth's. We take our gravity for granted; we have been born and have lived all our lives in a one-gravity field. One "gee" feels perfectly normal to us. But not to a whale. Whales are mammals, just as we are. They breathe air, just as we do. But they live in the virtually weightless environment of the sea, where their massive bodies are buoyed up by the water. A beached whale usually dies of suffocation because it hasn't got the strength to expand its lungs once it must fight the one-gee pull we take so much for granted.

Lifting anything — a pencil, a person, a hydrogen bomb, a Solar Power Satellite — against one gee takes *work;* the more massive the payload, the more work required. The more work required, the more energy must be expended. Energy costs money. Every time we boost a payload into LEO, we do the same amount of work as hauling that payload up a mountain that is 4000 miles high. Most engineers use a slightly different analogy: Boosting a payload into LEO is equivalent, in terms of energy spent, to climbing out of a hole 4000 miles deep. Space engineers speak of Earth's gravity field as a *gravitational well.* Here on the surface of our planet we are standing at the bottom of the well. Up in orbit, a spacecraft is at the top of the well. It has cost us the same amount of energy to get that spacecraft into orbit as it would cost to lift the same payload 4000 miles straight up.

Once we have climbed out of the gravitational well and established even a low orbit around the Earth, it costs very little more in energy to go anywhere else we desire within the Solar

System. The Moon is still a quarter of a million miles away, but in terms of energy expenditure (and therefore dollars) it costs less to go from LEO to the Moon than it does to climb up from Earth's surface to LEO. Far less. That first step is a bitch. A spacecraft must achieve a velocity of at least 18,000 miles per hour to establish itself in LEO. But to go from LEO to an orbit around the Moon costs only an additional 8000 mph worth of velocity. For another 5500 mph the craft can land softly on the lunar surface. More than half the energy needed to go from here to a soft landing on the Moon is spent clambering out of Earth's gravitational well and establishing a low earth orbit. Think of that: more than half the energy, to travel only the first hundred miles on a quarter-million-mile journey! Clearly, anything that can be done to reduce the cost of that first step will greatly reduce the cost of all space operations that originate on Earth.

When I worked on the Vanguard Project, it cost roughly $5000 to put a pound of payload into orbit. The fare for a man of my weight, using Vanguard economics, would have been $875,000 for a ride into low earth orbit. The earliest Space Shuttle missions promise a cost per pound of payload delivered into LEO of about $325. My ticket price for an early Shuttle flight will be $56,875 — if I can keep my weight down to 175 pounds. But that is only for the earliest Shuttle flights.

In *The Third Industrial Revolution* (Ace Books, 1979), engineer/author G. Harry Stine points out that the cost of flying into orbit aboard the Shuttle could come down to $50 per pound toward the end of the 1980s. Such cost reduction would come from gradual improvements in the Shuttle's performance: nothing terribly new or dramatic, merely the kind of capability upgrading that aircraft manufacturers provide routinely for commercial aircraft. Stine compares this to the cost of a traveling salesman who spends 18¢ per mile for automobile travel, plus $25 a day for meals and hotel expenses (both ridiculously low in today's inflated economy), and is on the road 250 days per year. This computes to a cost per pound of salesman (assuming 170 pounds) of slightly more than $60, of which transportation costs are $26.47 per pound. At a cost of $50 per pound of payload in LEO, the economics begin to look very bright for large-scale operations in space. And as Stine and many others have shown, the Shuttle is only the *first* reusable spacecraft to be flown.

Improvements in Shuttle design and totally new ideas that are currently little more than laboratory experiments will reduce the price of going into orbit just as surely as the price of electronic computers has plunged from millions of dollars in the 1960s to mere hundreds today.

The Space Shuttle is the first of its kind, a hybrid type of vehicle: part rocket, part aircraft. Engineers are fond of describing the camel as a horse that was designed by a committee. NASA's Space Shuttle was designed to satisfy a welter of conflicting needs. The result is a sort of aerospace camel that is terribly restricted in what it can accomplish.

It will greatly reduce the cost of getting into LEO, true enough, but not as dramatically as later space-planes will. And LEO is the first and last port of call for the Shuttle. It cannot — it is not designed to — reach the higher, more useful orbits.

If you want to place a communications satellite into geosynchronous orbit, 22,300 miles above the Equator, the Shuttle cannot do the job by itself. It must carry along an extra rocket engine system in the payload bay, which, after the Shuttle has attained LEO, will boost the comsat onward to GEO.

The entire Shuttle program has been a sobering microcosm of how NASA was maltreated throughout the 1970s by the White House, by the Congress, and — at bottom — by the American public: you and me. NASA's original plans for the Shuttle called for a two-stage vehicle that would be entirely reusable. It would have looked like a pair of sleek airplanes, one riding atop the other, rather like the Shuttle orbiter when it rides atop its carrier Boeing 747.

The big carrier was to lift the smaller orbiter up to the topmost fringes of the atmosphere, some 20 to 40 miles high, and then release the orbiter and fly back to land at an airfield. Next, the orbiter would proceed into orbit, accomplish its mission there, and then re-enter the atmosphere and land like an airplane, just as today's orbiter does. Both stages would be manned, both would be "fly-back" vehicles and completely reusable. But that, decided the cost-conscious cretins of Congress and the Office of Management and Budget, would be too expensive. They trimmed NASA's development funds down below the minimum needed for a fully reusable Shuttle, forcing the engineers to design a hybrid, "stage-and-a-half" Shuttle that is much less useful than it could have been.

In the late 1960s and early '70s, while Apollo astronauts were exploring the Moon and American prestige was literally sky-high, Washington quietly cut the heart out of all space projects. The exploration of the Moon was halted so abruptly that NASA did not even have the funds to continue receiving data from the automated instruments left on the lunar surface by the Apollo astronauts. The instruments were turned off while they were still functioning and sending valuable data to us.

Only one new program was initiated, although the Nixon White House brought it forth with great fanfare: the Space Shuttle. It was a sound strategic idea, to devote the major effort of our space program to a transportation system that would reduce the cost of getting into orbit, thereby opening the door to all sorts of future space operations. But like the equally hyped "War on Cancer," the tactics used to implement the Shuttle program were shoddy. From the very beginning, Congress and TOMB consistently allocated less money than NASA's studies showed would be needed to design and build the Shuttle.

It reminds me of the old Italian story about the farmer who decided that his mule ate too much hay. He cut the animal's feed ration in half. The mule continued to work as stolidly as always. So the farmer halved its hay ration again. The mule's ribs began to show, but it kept on working faithfully. After several more cuts in the feed allotment, the farmer stopped feeding the mule altogether. The mule kept on working for another day or so, then keeled over and died.

The farmer grew furious and began kicking the emaciated carcass, crying loudly, "Ingrate! Here I teach you how to work without eating and you go and die on me!"

The cuts in space budgets of the 1970s forced NASA to make a very tough decision. Quite clearly, the Agency's options were either to try to develop a Shuttle on the shoestring being offered or to close up shop altogether. NASA took the shoestring, and damned near strangled on it.

In his book *Enterprise* (William Morrow & Co., Inc., 1979), Jerry Grey, a professor of aerospace sciences at Princeton University and one of the most active consultants in the industry, described the political situation in the early 1970s this way:

A war-weary, inflation-scarred Congress, besieged by voluble lobbyists and constituents clamoring on all sides for desperately needed at-home social reforms, quickly developed an almost knee-jerk negative response to anything so costly and politically unpopular as an extensive manned-space-flight program.

In previous years, a group of liberal senators — Walter Mondale, Clifford Case, Jacob Javits, J. W. Fulbright, William Proxmire, George McGovern, Edmund Muskie, Birch Bayh, and Edward Kennedy — had made a practice of opposing space budgets mainly as a gesture to demonstrate their liberal views to their constituents . . .

This was the tragedy of the 1970s, the equating of liberalism with antispace, the *either/or* fallacy that said you can be *either* for social welfare *or* for a strong space program, but not for both. This narrow vision permitted the murder of the Apollo program and the strangulation of almost everything else we stood to accomplish in space.

Faced with an increasingly hostile Congress and a barricaded White House, NASA's administrators soon realized that they would not be allowed to develop the Shuttle that they felt was

needed to do the job: a fully reusable two-stage vehicle that
would be the most efficient and economical space transportation system the human mind could create.

Like the man who couldn't be near the woman he loved,
NASA decided to love the one it was near. The Shuttle NASA
wanted to build was beyond reach, so the engineers started to
design a Shuttle they could afford. Even at that, the funds actually appropriated by the Congress, year after year, were considerably less than the amounts really needed. NASA kept
doggedly at work, like the farmer's mule, hoping that they
could find a way to do the job within the strangling budget
limits or, if not, begging Congress for more money. This is called
"buying in." It is often done by a private contractor who, knowing that the job he wants to take on will cost more than the
government accountants have allowed, accepts the contract for
the low figure and comes back later for the money to finish the
job.

Cost-conscious politicians like Senator William Proxmire (D.,
Wisc.) have made much of contractor "overruns" in regard to
NASA and Defense Department contracts. In the aerospace
industry there is a bitter *bon mot* known as Cheops' Law: Nobody ever delivers on time and within budget.

But cost overruns are often built into a project at the very
beginning by unrealistically low cost limits imposed by some
bean-counting administrator who wants to look good in the
bureaucratic pecking order. ("See, I can get it for you whole-
sale!") It's as if you call in a carpenter to build a set of bookshelves in your living room and you tell him, at the outset, that
you wish to spend no more than $100 on the job. The carpenter
takes a look at what is actually needed and estimates that the
lumber alone will cost $75, the other materials at least another
$20, his own time will cost $50 or more. That's $145, minimum.
If the carpenter isn't particularly hungry, he will give you a $145
estimate, take it or leave it. But what if you are the only person
in town who hires carpenters? The only employer. The only
source of income for carpenters. Then perhaps he will reluc-

tantly agree to the $100 figure and either deliver exactly $100 worth of work — leaving you with a shoddy or even uncompleted set of bookshelves — or he will come back to you while the project is only partially completed and ask for more money. During the fifteen years I spent in the aerospace industry I saw more cost overruns generated that way, by the government's arbitrary underestimation of a project's cost at the outset, than for any other reason.

Certainly the aerospace corporations, large and small, were out to make profits for themselves. Certainly there were cases where poor management or sloppy performance made costs skyrocket. Ironically, when the situation gets so bad that the corporation looks as if it is about to collapse, the government steps in with mammoth financial aid. That is what happened to Lockheed in the 1970s and, in the private sector, to Chrysler in the 1980s.

Government agencies "buy in" to programs that they know are underfunded from the outset, because they want to do those programs, for one reason or another. Bureaucracies, just like any biological organism, have a life force to them, and they will do whatever they feel is necessary to ensure their own survival. NASA realized that without a Shuttle program, it was doomed as a viable agency. Already, thanks to the murder of Apollo, thousands of highly skilled men and women were leaving the Agency for jobs elsewhere — or were forced onto the unemployment rolls. One very bright physicist I knew in Boston, after more than a year of unemployment, told friends he was "on a Nixon fellowship."

NASA had no choice but to accept the Shuttle program on Congress's terms, like it or not. So the engineers went back to their drawing boards, literally, and started to design a Shuttle that would fit within Washington's meager pocketbook. They reluctantly went to the "stage-and-a-half" design, where the Shuttle takes off like a rocket, sheds it solid-propellant rockets and propellant tank, and then the remaining orbiter lands like an airplane.

The orbiter is actually a glider, because when it starts back into Earth's atmosphere it has no more propellants left for its engines. Imagine flying an aircraft "dead stick" from 200 miles' altitude and a speed of some 18,000 mph to a safe, airplanelike landing on a runway in California.

I tried it. I flew the Shuttle simulator, the computerized mockup in the Johnson Space Center, near Houston. The cockpit controls are all "live," but linked to a computer feedback system rather than to an actual flying Shuttle.

The orbiter flies somewhat like a hot brick; it was very nose-heavy when my clumsy hands were on the controls. While the instruments told me that we were pulling four gees and the plane was rolling over on its back, and the technicians who ran the simulator were gleefully informing me that I'd just ripped the wings off, I took my hands off the controls. And the orbiter's computer automatically got us back on course for a safe landing. All I had to do from that point on was to press the button that lowered the landing gear.

In a real flight we would have burned up in the atmosphere like a falling star or broken apart because of the excessive gee load.

In spite of all the delays and all the problems, the first orbital test flight of the Shuttle was picture-perfect. With astronauts John Young and Roger Crippen at the controls, the *Columbia* took off from Kennedy Space Center on April 12, 1981 — exactly twenty years after Yuri Gagarin made the first manned flight into space. After 36 orbits of Earth, *Columbia* landed at Edwards Air Force Base to the rejoicing of more than 100,000 men, women, and children who had camped out all night on the California desert to see history made before their eyes.

Amidst all the jubilation over the successful test flight, doughty Chris Kraft summed up the whole Shuttle experience with one sardonic line, "We just became infinitely smarter." But I cannot remember a single test program of a hot new airplane in which there wasn't at least one serious accident. Even with the spectacular success of *Columbia*'s first flight, if a multibillion-dollar Shuttle crashes, it could mean the end of the American manned space program. That's how close to NASA's jugular the knife is. That's why the Shuttle astronauts spent thousands of hours in the simulators, learning all there was to learn about flying the bird before lifting off from Cape Canaveral.

While the "stage-and-a-half" design was being drawn, NASA went out to seek customers for its Shuttle, organizations to provide the payloads that the Shuttle would take into orbit. This is sound marketing strategy. The rationale for building the Shuttle was based on the belief that there would be so much traffic wanting to get into orbit that an economical Earth-to-LEO transportation system would be sorely needed in the 1980s. So NASA went out to find who these customers might be.

Major industrial corporations and many smaller companies signed up for Shuttle flights. They had ideas about new industrial processes, research problems, and materials tests that they wanted to do in the zero gravity and high vacuum of orbit. NASA offered to rent them space aboard the Shuttle's capacious cargo bay.

NASA became sufficiently "hip" in its marketing and public-relations efforts to come up with the idea of a "Getaway Special": on some of the Shuttle's earliest flights, NASA will rent to individual persons or organizations five cubic feet of cargo bay space, for whatever experiment they want to put into it.

The cost of a Getaway Special five-cubic-foot space is $3000. The cost of renting the entire payload bay of a Shuttle flight runs up to $20 million or more: expensive, but considerably lower than corporations now pay for the launch of a relatively small communications satellite. One Shuttle launch could insert several such satellites into orbit, and even retrieve old ones for return to Earth.

NASA is also offering space aboard the Shuttle for valid college and high school student experiments, for free, as part of its long-standing commitment to further education in the United States.

But despite all these markets, NASA quickly found that its biggest potential customer was the US Air Force. The Department of Defense has awesome strength within Washington's corridors of power. Routinely, DOD's budget is so huge that it can spend as much as NASA gets annually in two weeks or less.

Back in the 1960s the Air Force had its own manned space program, but this was shut down (to the fury of some Pentagon brass), and sole responsibility for manned space flight was handed to NASA. Thus the Air Force and NASA had no great love for each other, as one bureaucracy to another. But by the early 1970s NASA needed the Air Force as a customer for the Shuttle. If the Air Force said it did not need a Space Shuttle, the chances were excellent that Congress would scrap the entire program and NASA would be virtually out of business.

The Air Force wanted *a* shuttle, to be sure. Its own, preferably. But since Washington would not allow that, the Air Force demanded that NASA make the Shuttle big enough to take on Air Force payloads, and maneuverable enough to land at any of several airfields, not just one that was preselected before takeoff. All of this meant that NASA had to go back to the drawing boards *again* to incorporate the Air Force's desires into the Shuttle design. Which NASA dutifully did.

Then the delays started. In their effort to stay within budget and on time, NASA took some daring shortcuts with the design of the new main engines for the Shuttle. Engine design is still something of an art, and the shortcuts resulted in failures. Engines burned out, engine tests were shut down prematurely, engines did not perform as calculated. The bugs were worked out, as they always are. But it took time. And money.

The new re-entry heat protection system, the heat shield, also caused untold grief. Instead of one single re-entry heat shield, as was built for the simpler Apollo and earlier manned spacecraft, the Shuttle heat protection system consists of thousands of brick-sized thermal tiles glued onto the belly, nose, wings, and flanks of the orbiter. They are astounding things, these ceramic tiles. Thanks to modern materials technology, you can heat one end of a tile with a blowtorch while holding its other end in your bare hand. But they refused to stay glued to the orbiter! The tiles kept dropping off. The technicians kept gluing them back on, muttering darkly about "leprosy." Re-entering the atmosphere at 18,000 mph with holes in your heat shield is not the best way to save money. Or lives.

That problem was solved at last, but only after more delays and expense for an already-troubled Shuttle program.

Critics of the Shuttle say that it has become just as expensive as the older, throwaway boosters it was supposed to replace. The European Space Agency is developing a throwaway booster called Ariane, which, they announced, would be lofting payloads into orbit sooner and more cheaply than the Shuttle. Unfortunately, on its second test, an Ariane booster crashed into the Pacific. ESA has hushed, somewhat.

The long delay of the Shuttle's first flight spurred many of NASA's erstwhile customers to seek other boosters for their payloads. Commercial operations such as communications satellites and even government operations such as weather and observation satellites could not wait indefinitely while the Shuttle program dragged on. So many potential Shuttle payloads were switched to such well-proven boosters as the Delta and Titan.

Some antimilitary critics insist that the Shuttle is little more than a thinly disguised Air Force vehicle. It is quite true that a large percentage of the Shuttle's flights will be Air Force missions, launched not from Kennedy Space Center but from Vandenberg Air Force Base, in California. An Air Force program office is now set up at Johnson Space Center, headquarters for the Shuttle effort. And about half of the 100 astronauts training for the Shuttle are military officers.

But this does not negate the fact that the Shuttle will boost hundreds of payloads for civilian customers. Private companies like Boeing and several airlines have already expressed interest in taking over Shuttle operations once the system has been thoroughly tested. They want to run a commercial "space line" between the ground and LEO; they believe it could be a profitable business.

Certainly, once the Shuttle becomes operational, the question of NASA's role arises. Should NASA operate a transportation service, or should the Agency remain in the R&D role for which it was originally created? Should the operational Shuttle system be turned over to another part of the government, for example, the Department of Transportation, or should an airline or other private firm be allowed to buy the equipment and set up profit-making operations?

Some friends of the Shuttle feel that this kind of speculation is not only premature, it is harmful. They fear that the Prometheans may be promising too much too soon, and the realities of the Shuttle's long and perhaps difficult testing period may disillusion the public and, even more important, the Congress. Certainly, if the Shuttle turns out to be plagued with troubles during its flight testing program, or if one of the vehicles crashes, all the bright promise of the future is clouded over.

But clouds do not last forever. The first manned Apollo craft suffered a disastrous fire during a ground test that killed astronauts Gus Grissom, Edward White, and Roger Chafee. The Apollo program was set back 18 months while the cause of the fire was determined and protective measures designed for subsequent spacecraft. We buried our dead, and kept moving forward.

The simple truth is that there *are* customers, civilian and military, and they will buy the cheapest and easiest transportation available to take them into orbit. A study released early in 1981 by the American Institute of Aeronautics and Astronautics (AIAA) shows that the commercial market for satellite launch vehicles is easily double the capability of the four Shuttles now being built by the government. And the Russians are flight testing a reusable, delta-winged shuttle craft, according to aerospace sources.

Back in the early 1930s the Douglas Company built a sleek

two-engined all-metal airplane aimed at making commercial air service a financial success. That plane, the DC-2, was *almost* right for the job. The experience gained by designing and flying it gave Douglas's engineers everything they needed to produce the DC-3. The DC-3 was not a lot bigger or faster than the DC-2. But it was big enough, fast enough, carried enough passengers in enough comfort and safety to make the airline business profitable. For three decades, the DC-3 was the bellwether of airlines all around the world.

The argument about today's Space Shuttle is really a debate over whether it is a DC-2 or a DC-3. Most aerospace engineers feel that it is a DC-2. The 1980s will be spent checking it out, using it for the commercial and military missions that are already "booked" for orbit, and learning what is needed to upgrade it into a DC-3.

The 1990s will see the burgeoning of a huge commercial market in space. And a heavy military presence in space, as well. The Shuttle will be the key to taking that all-important first step toward LEO.

All of NASA's eggs are in the Shuttle basket. To a huge extent, all of our hopes for a strong space program are in there too.

23

The Miners' Bonanza

Fire is the test of gold; adversity, of strong
men.

—SENECA

Everything said here about the new space program, especially
the economic forecasts, is based on a false assumption. The tacit
assumption has been made that all the materials for everything
done in space must be lifted up from the surface of the Earth.
That's doing things the hard way, boosting up vast tonnages of
hardware, supplies, and people from the bottom of a 4000-mile-
deep gravitational well. There are plenty of other sources of
raw materials and supplies in space, waiting for us.

If there is one thing that our space explorations have already
shown us, it is that the Solar System is incredibly rich in natural
resources. The raw materials for multibillion-dollar industries
lie strewn around the large and small bodies of the Solar System, waiting for us to pick them up and use them. Although
most of these raw materials lie millions of miles from Earth, it
will actually be cheaper and easier to get them and use them
for space industries than to bring up raw materials from Earth.
Remember, space is not a barrier: It is an open highway. Just

as the Europeans of the fifteenth century turned the world's oceans into avenues of commerce, so will we turn the dark void of space into golden trade lanes of worldwide prosperity.

We know that the Moon, a "scant" quarter-million miles away, contains plenty of valuable construction materials: aluminum, titanium, carbon, silicon, oxygen, and others. Aluminum and titanium can be used to build structures on the Moon or elsewhere in space. Silicon and carbon are vital ingredients in solar cells and in manufacturing electronics components like transistors. Oxygen is indispensable for life support, and when burned with aluminum powder makes a fine, cheap rocket propellant. All of these materials are abundant on Earth, of course. But it takes vastly more energy, and money, to lift them up from Earth's surface. Manufacturing in space will be much cheaper if lunar raw materials are used instead of terrestrial raw materials.

Lunar mines will not be deep pits that leave huge scars on the Moon's surface. A couple of bulldozers scraping up a foot or two of dusty "topsoil" across an area the size of two football fields could provide about a million tons of ore per year quite easily.

We know comparatively little about the subsurface composition of the Moon. While some of the rocks brought back to Earth by the Apollo astronauts may have originated deep below the surface and been blown out by volcanic eruptions or meteor impacts, we have little hard data on anything except the surface.

We know from astronomical measurements that the Moon's density is 3.4 grams per cubic centimeter, which is much lighter than Earth's 5.2 (water's density is fixed at one gram per cc). This leads to the conclusion that the Moon is relatively poor in the heavier metals, such as iron, gold, uranium, and so on. The Apollo samples seem to support this conclusion: They contain light metals only.

However, the big flat "seas" of the Moon — the maria — are very probably sites of titanic meteor impacts, billions of years ago. Satellites orbiting the Moon have detected anomalous concentrations of mass beneath the maria: *mascons,* in NASA jargon. Heavy material is buried under the maria, most likely the shattered remains of the meteors that caused these vast circular impact scars. If there are iron, nickel, gold, or platinum lodes on the Moon, the maria are where they will be.

The Moon's low gravity and airlessness make it cheap and easy to launch payloads from its surface. We won't need rockets for the task; an electric catapult a few miles long will do the job nicely.

Arthur C. Clarke suggested catapulting payloads off the Moon back in the 1950s. He correctly saw that the Moon, with a surface gravity only one-sixth that of the Earth and an atmosphere so thin that it's better than high-priced vacuums on Earth, would allow engineers to use catapults driven by electricity derived from solar cells or nuclear reactors.

Since the Moon turns on its axis at exactly the same 27-to-29-day period it orbits around the Earth, the Moon keeps the same face always pointing Earthward. There is no "dark side" of the Moon; both sides receive the same amount of sunshine each month. Indeed, a "new moon" occurs when the Earthside of the Moon is totally in darkness while the Farside is totally in sunlight.

A lunar "day" is about 14 Earthdays long. Solar panels could generate electricity without interruption for 14 days straight, while additional panels produced more electricity to charge storage batteries for the 14-day-long lunar night. The panels themselves and most of their associated equipment could be built from lunar materials scooped from the surface and processed by fairly standard equipment brought up from Earth.

Alternatively, nuclear reactors could be hauled up from Earth to provide constant electric power no matter what the day/night cycle.

Clarke's concept of a catapult has evolved over the years into the "mass driver" of Gerard K. O'Neill, who originally invented the device for quite a different purpose: to accelerate subatomic particles for laboratory physics experiments. O'Neill is a professor of physics at Princeton University and the originator of the modern concept of building massive colonies in space between the Earth and the Moon (see Chapter 28).

When he began thinking about habitats in space big enough to house thousands of permanent colonists, he quickly realized that the only economical way to build such mammoth structures was to take the needed raw materials from the Moon. Thus he hit upon the idea of adapting his laboratory accelerator into a miles-long lunar mass driver that would hurl bucketloads of lunar ores out toward the refineries and processing plants waiting hungrily in space where the colonies were to be built.

As O'Neill envisions it, the mass driver would look like a long looped track, along which travel buckets big enough to hold a few hundred pounds of raw ores. The buckets are accelerated along the track by electrical energy, and as they reach the lunar escape velocity of 1.5 miles per second (nearly 5400 miles per hour) the payloads are detached and flung off into space. The buckets and every other part of the mass driver remain in place to be used over and over again, year in and year out. It is a very efficient system, which means it will cost little to boost lunar ores off the Moon's surface.

A mass driver of a few miles' length must accelerate its payloads to nearly 30 gees: 30 times the force of gravity that we experience on Earth. That kind of acceleration is for inanimate payloads only! Anything fragile, such as a human being, would be crushed to a pulp. O'Neill's research has shown that mass drivers could attain accelerations of hundreds of gees, which means they could be made shorter, and therefore cost less to build.

Human travelers could be catapulted off the Moon, but a mass driver for people would have to be at least ten miles long. The extra length is needed to keep acceleration down to a level the human body can tolerate. With aluminum and oxygen so plentiful on the Moon, it is likely that humans will go to and from the Moon in cheap alumi-oxy rockets. (The way God meant for people to travel!)

The Moon is poor, indeed barren, in the most precious re-source to be found in space: water. Every indication is that the Moon is totally dry. The chances of *manufacturing* water on the Moon are remote, but not altogether impossible. The problem illustrates beautifully how we must adjust our way of thinking once we begin living and working off the planet Earth.

On Earth, energy is expensive and water is cheap, relatively speaking. On the Moon, energy from sunlight could be literally dirt cheap, but water will probably be the most expensive commodity to be had.

David R. Criswell, of NASA's Lunar Science Institute, has found through microscopic study of lunar samples returned by the Apollo astronauts that, even though the Moon seems water-less, there is some hydrogen gas imbedded in the lunar soil. Presumably the hydrogen and other light gas atoms are trapped in the soil when they waft in on the Solar Wind that blows through interplanetary space. It may just be possible to com-bine oxygen from the soil with this trapped hydrogen — or with the ephemerally thin hydrogen gas in the Solar Wind — to literally manufacture water on the Moon. Considering that the Solar Wind is so tenuous that a volume of space the size of the entire Earth may contain less than a hundred pounds of hydrogen, we can visualize the immensity of the problem. It might be better to look for a "waterhole" elsewhere.

We know that hauling water up from Earth is the most expen-sive way to get it. To use that 4000-mile-deep gravity well as a water well makes no sense unless there is no possible alterna-tive. Where is the next nearest place to obtain water? The planet Mars.

Thanks to our Viking spacecraft we know that the shining white caps on Mars' poles are composed of frozen water. Before the Viking landings, most astronomers assumed that the Mar-tian polar caps were made of frozen carbon dioxide: dry ice. But they are water. Mars orbits tens of millions of miles away from the Earth and the Moon, and it may at first glance seem ludicrous to go that far for water when we can haul it up from the Earth itself. But Mars is a small planet, with scarcely a third of the gravitational pull that Earth has. Traveling millions of miles in space is merely a matter of time. If we are willing to spend the transit time necessary to travel between the Earth/Moon system and Mars, it will be cheaper to get water from Mars than from Earth.

Viking's landers told us something else of inestimable value about Mars: Human explorers can live on the Martian surface indefinitely, using water from the polar caps and taking oxygen from the thin Martian atmosphere for their life support. They may even be able to grow food crops in the soil of Mars.

Mars is a cold, seemingly barren world. The daily weather reports telemetered from the Viking landers show typical overnight lows of 120 degrees below zero, or worse, even in summer. The air is almost pure carbon dioxide and thinner than the air of Earth a hundred thousand feet above the tip of Mt. Everest.

Yet there is oxygen and even nitrogen in that wisp of an atmosphere. Human explorers could set up a Buckminster Fuller–type geodesic dome on Mars' surface, pump it full of oxygen and nitrogen extracted selectively from the Martian atmosphere, and create an Earthlike air inside their dome. With water from the poles they could live within the dome, and perhaps even grow food crops to become self-sufficient.

Scientists want to study Mars to see if there is life hidden beneath its red sands. The Viking landers' automated equipment detected strange chemical reactions in the soil, neither ordinary terrestrial chemistry nor anything recognizable as biology. No one can say that life definitely exists on Mars, but the case is not yet closed.

The men and women who mine the Moon and staff the factories and research stations in orbit around the Earth will look to Mars as their "waterhole." The scientists who want to explore Mars could finance their expeditions by selling water to the miners and factory workers of the Earth/Moon system.

But the real miners' bonanza lies even farther out than Mars. *Trillions* of chunks of rock and metal are floating through space, for the most part between the orbits of Mars and Jupiter, some three times farther from the Sun than our own planet — several hundred million miles from Earth. They are called the asteroids, because to the astronomers who first discovered them they appeared as tiny stars in their telescopes. More properly, they should be called planetoids, because that is what they are: tiny planets, little chunks of solid matter.

The largest of them, Ceres, is some 500 miles in diameter. The smallest are probably no bigger than grains of sand. Some of them, such as Apollo and Icarus, swing in their orbits so close to Earth that astronomers can see they are not circular in shape. Apollo measures some 22 miles long and 10 across; Icarus is not even one mile wide.

The asteroids have been called "mountains floating free in space." They may be the remains of a planet that once orbited between Mars and Jupiter and somehow exploded. More likely, they are fragments that never coalesced into a planet, presumably because of the powerful gravitational tides from mammoth Jupiter, largest of all the Solar System's worlds.

Like mountains on Earth, the asteroids contain treasure-troves of metals and minerals. "Thar's gold in them thar hills!" Gold, silver, platinum, iron, nickel, copper, manganese, carbon, potassium, phosphorus, rare earths, water, organic chemicals — all the metals and minerals that the human race needs to survive and flourish — exist in the asteroids in megaton quantities.

Clark R. Chapman, planetologist at the Planetary Science Institute in Tucson, said in his book, *The Inner Planets* (Scribner's, 1977):

> But 30 or 40 [asteroids] exceed 200 kilometers, the length of Massachusetts excluding Cape Cod, and roughly 3000 exceed 20 kilometers (the size of Lake Tahoe). Literally trillions of uncharted boulders the size of a basketball or larger exist. For each asteroid there are 10 others one-third its size . . .

Consider: In 1977 the world's gold production amounted to slightly less than 1000 tons, for a total value of $1.3 billion. A single rocky asteroid of the type that astronomers call *carbonaceous chondrite*, no larger than 100 yards across (the length of a football field) could contain some $15 million in gold. As an impurity. The real value of the asteroid would be in the organic chemicals it contained. One asteroid of modest size. Imagine what asteroid mining is going to do to the price of gold.

Consider: A 100-yard-wide asteroid of the nickel-iron variety contains nearly four million tons of high-grade nickel steel, which is worth more than a billion dollars in today's steel industry. World production of steel averages less than 750 million tons per year, and experts forecast that there will be a world shortage of steel in the 1990s, if current trends continue. Yet there are hundreds of billion of tons of steel in the asteroid belt, waiting for us.

Bonanza, indeed.

These are not blue-sky pipedreams (or SWAG — scientific wild-ass guesses — as they are called in the aerospace industry). We have actual laboratory data on the composition of the asteroids. Each day, upwards of 10,000 tons' worth of them fall into Earth's atmosphere. We call them "shooting stars"; astronomers call them meteors. Most of them are smaller than dust particles, and when they hit the atmosphere they burn up. But some meteoroids are big enough to reach the ground, and the surviving chunks are called meteorites by the astronomers. They are believed to be fugitives from the Asteroid Belt that have wandered close enough to Earth to be caught in our planet's gravity well and slide down to a fiery entry into our skies. By examining the meteorites we know much about the composition of the asteroids; the two are presumed to be one and the same.

Prospectors and miners will journey out beyond Mars, to the Asteroid Belt, armed with electronic gear and lasers, riding silent electrical rockets rather than braying burros. They will be followed by huge factory ships, sailing the placid sea of vacuum for years at a time, scooping in thousands of millions of tons of rock and metals, processing them into finished products on the long voyage home. If the history of industrialization on Earth means anything, the factories will inevitably be sited as close to the sources of raw materials as possible. It is always cheaper to transport finished products than raw materials. That's why Pittsburgh is where it is. And Birmingham. And Essen.

Like the whalers of New Bedford and the iron crews who manned the wooden ships of old, the men and women who journey out to the asteroids will be prepared to spend several years "before the mast." And like the whalers and sailors of those earlier centuries, they will share in the profits of their voyages and return to Earth richer than any corsair who ever waylaid a treasure galleon.

Even farther will they roam. To the comets that periodically swing past Earth's orbit, rich in organic chemicals and compounds and possibly holding the secrets of the Solar System's origin. To the myriad moons of giant Jupiter, where frozen water and rare isotopes of hydrogen and helium — fuel for nuclear fusion reactors — lie in abundance.

The twentieth century has seen the grandest voyages of exploration ever undertaken by humankind: Apollo, Viking, Voyager, Pioneer. The twenty-first century will reap the harvest of those explorations, a harvest of metals and minerals so abundant that scarcity will be a word used only by historians.

24

The Meek Shall Inherit . . .

> And death shall be no more; death, thou
> shalt die.
> —JOHN DONNE

It may well be that the first people to live permanently in space will be the old, the weak, the infirm. The usual vision of astronauts and spacefarers is that of young, vigorous people, the jut-jawed heroes of science fiction with their seductive movie starlets at their sides. Certainly the first astronauts and cosmonauts were men in the prime of their lives. Not the twenty-year-olds that science fiction expected, but experienced jet fliers and test pilots in their thirties and forties, knowledgeable and cool under stress.

NASA and the American media made the mistake of presenting the astronauts to the public as demigods — part superhuman hero, part superintelligent computer, all-American and all family men. As Tom Wolfe showed in his excellent book *The Right Stuff* (Farrar, Straus, & Giroux, Inc., 1978), the astronauts' true personalities were just as diverse as any group of greatly accomplished men. But they were, on average, more like Clark Gable and Spencer Tracy's portrayals of test pilots than the stiffish, apple-pie image created by NASA and the media.

In the Soviet Union, cosmonauts are adored with an intensity reserved in the West for rock stars. Probably because there are so few outlets for hero-worship in the socialist "workers' paradise," cosmonauts are mobbed by screaming teen-agers wherever they go in public. And why not? They are the embodiment of Soviet ideals: dedicated, highly trained men who use the highest technology that the human mind has created to extend the domain of the human race. Besides, they are somewhat younger than American astronauts, on the average; and according to teen-aged Russian girls, most of them are dashingly handsome.

Although several women are in astronaut training for the Space Shuttle program, to date only one woman has even been in space. She is Valentina Vladimirovna Tereshkova, who, at the age of 26, piloted the Voshkod 6 one-seat spacecraft through 48 orbits of the Earth, in June 1963. She was in space for nearly 72 hours. Cosmonaut Valery Bykovsky orbited near her craft in Voshkod 5. Tereshkova married cosmonaut Andrian Nikolayev in November 1963 and has retired from the Soviet space program. She has one child now and shuns the public limelight.

Before Yuri Gagarin made the first manned space flight, April 12, 1961, many people believed that humans could not survive in space. Weightlessness, loneliness, psychological and physical problems would forever prevent humans from flying in space, they thought. Twenty years of human space flight have shown that not only were these pessimistic guesses wrong, but that there are psychological and physical *benefits* to living in a zero-gee environment. The biggest problems foreseen by space physicians stemmed from the effects of weightlessness on the human body. Muscle tone would degrade in an environment where the muscles are not constantly stressed by the pull of gravity. This could have drastic consequences for the cardiovascular system. The heart, after all, is a pump made of muscle. If the heart muscle is weakened by prolonged exposure to zero gravity, what happens when the heart feels a full Earth gravity once more?

The earliest space flights also showed that calcium production in the bones tended to drop off as the astronauts lived in zero gravity. Would this make the bone structure too weak to allow an astronaut to stand up on Earth?

Weightlessness, the effect of being under zero gravity conditions, is also called "free fall." This is because a satellite in orbit is, quite literally, falling toward the center of the Earth. Its huge forward speed (18,000 miles per hour at the lowest orbits) carries the falling satellite out beyond the curve of the Earth's bulk. It never crashes into the Earth, but it is continually falling.

The sensation of weightlessness, then, is something like the sensation you get when an elevator starts to drop. But it's not just a momentary lurch in the pit of the stomach. It's continuous; eternal. As Isaac Asimov said of his first ocean voyage, "It came as a shock to realize that when I wanted to sleep at night, they didn't turn the ocean off."

Some astronauts and cosmonauts have become nauseated when they first encountered weightlessness. But apparently space sickness, like seasickness, is temporary. After a few hours the inner ear adjusts to weightlessness and the nausea disappears. NASA medics have found that training and psychological preparation also help to reduce the intensity and duration of space sickness.

As the astronauts who spent months at a time aboard Skylab reported, weightlessness can be fun. There are advantages to zero gravity. You can float effortlessly from one end of your spacecraft to another, maneuvering with fingertip touches against the bulkheads. You must be careful about shaving, eating, and housecleaning because whiskers, crumbs, and litter do not drop to the floor; they float in the air currents and inevitably end up in the filter screens over the air vents. On the other hand, adjusting to life back on Earth can have its problems. One Skylab astronaut complained that it took him days to remember that if he released his ballpoint pen in midair, the damned thing would crash to the floor, instead of hanging there waiting for him to take it in his hand again.

The longest-duration flight in space, as of this writing, was the 185-day mission of Soviet cosmonauts Valeri Ryumin and Leonid Popov, aboard Salyut 6. Ryumin has now spent almost a full year in space. He was on the earlier, 175-day mission aboard Salyut 6 with Vladimir Lyakhov. Concerning that earlier flight, writer Nick Engler stated in *Omni* magazine:

> On this [six-month] mission . . . Ryumin and . . . Lyakhov achieved *homeostasis,* a healthy physiological balance with their space environment. Cardiovascular changes, muscle atrophy, loss of calcium from the bones — all the biological problems that have worried aerospace physicians — slowed and stabilized.

In other words, the human body adapts to the weightless environment. The problems that the medical people fretted over are problems of *returning* to a one-gee environment. There is no apparent physiological reason why human beings cannot live in space permanently, given permanent life-support equipment and supplies. In fact, some of the consequences of prolonged space flight are rather pleasant. The curve of the spine, bent by fighting Earth's gravity, unfolds a little; the spacefarer becomes an inch or two taller. The production of body fluids and minerals adjusts to lower levels because of the lessened stress on the organs and bones. There is a tendency to eat less because the body is doing less work and demands less fuel. Dieting is easier in space.

The major problem in space is an increase in radiation. On Earth our atmosphere and the geomagnetic field shield us from most of the radiation that floods outer space. In orbit, the shielding of the atmosphere is gone, and for spacecraft that venture out to the Moon or beyond, the protection of Earth's magnetic field is also left behind.

The structure of the spacecraft itself is enough of a radiation shield under ordinary circumstances. But if a solar flare should blast a radiation storm toward the spacecraft, the crew has only

two choices: return to Earth before the storm hits, or bring
along a thick-walled "storm cellar" where you can stay until the
storm blows by. The Apollo astronauts were prepared to abort
a mission and return to Earth if a solar flare erupted. Longer
missions beyond Earth's magnetic field will have to include
massive protection from cosmic radiation and solar radiation
storms.

The first astronauts and cosmonauts were jet jockeys, military
fliers, test pilots. Today, men and women who are not pilots are
being trained to work in orbit for weeks at a time, aboard the
Space Shuttle. In the near future, the Shuttle will carry media
reporters, VIP's, perhaps even science fiction writers into orbit.

The next generation of spacefarers will be the construction
crews and factory workers. And after them will come the first
permanent space residents: the elderly and infirm.

The construction crews will build the permanent space stations, the lunar mining facilities, the orbital factories. Like construction crews or oil riggers in any far-off, dangerous area, they
will work on contract for 30 days, 90 days, one year, and receive
hazard pay in addition to their normal wages. As much of their
work as possible will be automated, of course, but there is a
better-than-even chance that a few of the men working Alaska's
North Slope oil fields today will be wrestling with weightless
building beams in orbit before they retire.

In New York there is a tradition that Mohawk Indians take on
the toughest, tallest steel work on any new skyscraper being
built. Will a few Mohawks tackle the really "high steel" jobs of
building space stations and orbital factories? Will teen-agers
living in West Virginia mining country today be directing the
lunar mining operations of the next decade? The opportunity
is there. The possibilities are endless.

Certainly these construction workers and miners and factory
hands will be chosen for brains, good health, adaptability. They
will work in space for fixed lengths of time, then return to
Earth. Surely many of them will go into space more than once;
the pay, the adventure, the achievements will be just too high

to sample only once. Just as surely, some of them will die in space. Building, mining, even factory work are not without their hazards, and the slightest misjudgment could bring swift and certain death in the hostile environment of space.

But eventually some of these workers may decide that they don't want to return to Earth. They don't want to fight gravity again, go through the readjustment period, battle the crowds in the cities, worry about the weather and pollution and real estate values. Especially as they age, the comforts of zero gravity (or the one-sixth gravity of the Moon) may begin to outweigh the pull of home.

Gravity kills us, on Earth. It ages us, making our flesh sag and tiring our muscles. As our bones become brittle (perhaps from too much calcium production?) they snap more easily if we fall in this one-gee environment. Living in space may not make us immortal, but the indications are that it could lengthen our lifespan significantly. Consider the advantages of low-gravity living for the elderly and infirm. The heart need not work so hard, pumping weightless blood through the cardiovascular system. And since the entire body is under much less physical stress under zero- or low-gravity conditions, the heart's workload is eased even more. The bones and muscles need not support the weight that they must bear on Earth. And the calcium loss that so worried aerospace medics a decade ago may mean that aging bones will not get so brittle in low-gravity environments.

Because the life-support system in a space habitat must be maintained with scrupulous care, or the whole habitat will be wiped out, the environmental conditions in a space habitat should be much, much better than anywhere on Earth. Clean, carefully filtered air. No air pollution. No pollens. As a lifelong asthmatic, I find that attractive. People who are incapacitated by allergies on Earth may be liberated by living in space.

I am not talking about some orbital old-age home, where "Golden Agers" quietly drift away the last few years of their lives. What I'm talking about is *life extension* — of producing an environment where men and women can lead active, useful lives for years, decades, beyond the time that the tyranny of gravity would permit on Earth. In orbit, or in the gentle gravitational clasp of the Moon, men and women who die fighting gravity on Earth in their sixties or seventies could live on, alert, active, vigorous, productive. Who knows where the limits are? On Earth the maximum lifespan has averaged "threescore and ten" since Biblical times. In space — a century? Sixscore years? More?

In a novel of mine, *Millennium,* one of the minor characters is an aged Russian ballet master who is so beloved by the Russian people that the Soviet government sends him to their lunar colony when he becomes too feeble to live any longer on Earth. As that part of the story streamed out of the typewriter in front of my somewhat astonished eyes, it became clear that:

One: We must have gentler boosters than the ones we now use. No cardiac patient could survive the three-gee launch of the Space Shuttle. We need the advanced, all-reusable Shuttle, the one that NASA originally wanted to build.

Two: In the low gravity and clean environment of a space habitat, a weak, dying old man could rather quickly become strong enough to resume his life's work.

Three: In low-gravity space environments, size and weight will be scant impediment to aspiring dancers. I have always envied Fred Astaire's grace and style. Perhaps in zero gravity, I too can learn to float like a butterfly.

The husband-and-wife writing team of Spider and Jeanne Robinson produced the definitive work on dance in weightlessness, a novel titled *Stardance.* Jeanne is a dancer and choreographer, not a full-time writer, as Spider is. When the story was first serialized in *Analog* magazine, the readers' response was overwhelming. There are millions of frustrated ballerinas in the world, chained by gravity, who saw their hearts' desire in that story of weightless dance.

There is already an art gallery, of sorts, in orbit. At least two paintings by the Russian space artist, Andrei Sokolov, were sent up to the Salyut 6 space station, where they helped to brighten the station's functional interior. And what will sculpture become in the weightless, inherently three-dimensional medium of space? Will tomorrow's sculptors emigrate to orbit, where they can work on meteors and asteroids of sizes ranging from a few inches to a few miles across? Will the grandest sculpture museum of history float freely in vacuum, a few hundred miles overhead?

As we saw in Chapter 21, the idle rich will patronize orbital tourist hotels, transforming today's jet set into tomorrow's rocket set. They may be the first to try weightless dancing. Who knows what totally new art forms may be created under the new physical and psychological conditions of space living?

Remember, in weightlessness no one needs a facelift. Clothing styles may be the functional jumpsuits of the movies, at first, but soon enough both men and women will begin to want clothing that enhances the freedom of weightlessness, the flowing loose kinds of style that go with the euphoria of zero gravity.

What will sex be like in free fall? To those who enjoy waterbeds, think of a chamber in which all six surfaces are warmly padded, inside which you can float endlessly, and in which the slightest touch produces a gentle movement. The possibilities are mind-boggling.

The new frontier of space offers possibilities of inestimable value not merely to corporations and stockbrokers. Jobs, profits, balance sheets notwithstanding, the new experiences waiting for us in space will lead to new arts, new joys, an unlimited opportunity to expand our horizons, to create new modes of living, to extend our lifespans far beyond the limits of Earth. It may well be that the Gospel has it subtly wrong: The meek shall not inherit the Earth, but the universe in all its magnificence.

As television's famous Mr. Spock would put it, "Live long and prosper." In space.

25

The Military

> It is to our interest to see that we are strong
> . . . Weakness cannot cooperate with any-
> thing. Only strength can cooperate.
> —DWIGHT D. EISENHOWER

Yes, the same Dwight D. Eisenhower who, as President, warned against the encroaching power of the "military-industrial complex."

He was right both times.

The problem with either/or thinking is that it straitjackets our minds and limits the options we may choose. The United States does not have to be *either* an imperialistic tyrant *or* a helpless isolationist wimp. We must be strong if we wish to survive: strong morally, economically, politically — and militarily.

The Gospel says that where your purse is, there will be your heart also. A great deal of our national welfare already depends on communications satellites, weather satellites, Landsats and Seasats, and the early-warning satellites that discourage surprise missile attack.

As we place more and more of our national welfare in space,

as we reach out to tap the enormous wealth waiting there for us, as we place an increasing share of our vital communications links in orbiting satellites, even more of our "purse" will be in space.

Major Stanley G. Rosen, of the Air Force Space and Missile Systems Organization (SAMSO), put it this way:

> [Commercial] and economic dependence on space is also growing. Moreover, we realize that projected commercial investment in space power generation, manufacturing, and other economic activities will contribute directly and significantly to the national interest. Therefore, we must regard these assets as an integral part of the strategic balance and must pursue all necessary avenues to defend them . . .

We must be prepared to defend ourselves in space as well as on Earth. Even within the territorial confines of these United States, we do not leave vital communications centers unguarded. We protect our possessions overseas. We have seen what happens when that protection fails, as it did in our embassy in Tehran.

Most of us believe instinctively that outer space should not become an arena for military conflict. Perhaps it is because space seems nearer heaven, or because it is a totally new, unspoiled area, virginal and clean. We want space to be free of Earthly evils, so much so that we tend to reject thoughts of militarizing space and even regard those who speak about military necessities in space as warmongers.

Yet, as Patrick Henry said, "Gentlemen may cry peace, peace — but there is no peace."

There is no peace. Not even in orbital space. The naïve desire to keep space free of the military was already a hopeless dream in 1957, when the first Sputnik was launched.

The human race first stepped into space for military purposes with the V-2 rocket. Then the Soviet Union, followed by the United States, developed intercontinental ballistic missiles in the 1950s. ICBM's are space weapons, boosted from launch pad to target by rocket engines. Sputnik and the celebrated race to the Moon were merely public relations exercises aimed at bolstering national prestige, once East and West had achieved rough nuclear and ICBM parity.

All the dreamers, the poets, the starry-eyed kids, and science-fiction writers — they *followed* the military into space. They did not, could not, lead. Why not? Because no one on Earth pays out billions of dollars for dreams, or prophecy. No political leader, no taxpayer, would risk a red cent on "flying to the Moon" in 1950. I know. I was there.

But once our national security was threatened, it was billions for defense. Even today, the Department of Defense can spend more on space than NASA can. To those who decry this state of affairs as proof that "the military-industrial complex" is evil and too powerful, I say this: It's your money they're spending. If you had demanded a vigorous follow-up to our stunningly successful Apollo program, we would have had such a commanding lead in space technology that we could have created a *Pax Americana* in space. Now it is too late. To those who disbelieve that American dominance of space would have been peaceful and benign, I can only point to history: After World War II we offered to share our nuclear technology with the world, openly and freely, providing only that international safeguards were created to prevent the proliferation of nuclear weaponry. The Soviets scotched that initiative, preferring to develop their own nuclear arsenal rather than trust the United Nations.

In 1955, President Eisenhower proposed a dramatic "open skies" policy, in which the US offered to present a complete report of its military forces, together with an inspection system and a plan for worldwide disarmament. Eisenhower later wrote:

> I proposed that we agree that outer space should be used only for peaceful purposes . . . Both the Soviet Union and the United States are now using outer space for the testing of missiles designed for military purposes. The time to stop is now . . .

The militarization of space did not stop in 1955. A quarter-century later, the militarization of space is proceeding faster than ever. The burden of space defense falls on the US Air Force. General David C. Jones, when he was Air Force Chief of Staff, summarized our space policy: "The Air Force affirms that its responsibilities in space include the duty to protect the free use of space by providing needed defense capabilities . . . in accordance with national policy and international law."

Major Rosen, of SAMSO, has pointed out that the Air Force's mission in space involves the military situation here on Earth. "Whatever the nature of the satellite placed in orbit," he explains, "it is dedicated to one fundamental purpose — to improve the lot of the American fighting man, be he on land, at sea, or in the air . . . "

Rosen states that the first goal of space defense is to prevent war; failing that, early warning of attack; finally, execution of the battle — winning the war if it comes to actual fighting.

Military satellites provide the key to several major defense capabilities today. Communications satellites link ground commanders, ships at sea, aircraft aloft, with the chain of command and each other, all around the world. Observation satellites provide early warning of missile launches and troop movements. Navigation satellites yield pinpoint accuracy for ships, planes, and submarines. Weather satellites allow detailed forecasts for any place on the globe. These are entirely passive military roles: important, but not aggressive. That period of space defense is being eclipsed. Active, aggressive weaponry is on its way to orbit. Indeed, it is already there. The Soviets have developed a satellite killer system capable of destroying satellites in low orbits. In October 1978, Secretary of Defense Harold Brown announced that the Soviet system was operational.

The system is as simple as it is deadly. A killer satellite is launched into an orbit that matches the orbit of the target satellite. When the two are side-by-side, the killer detonates its explosive warhead, destroying both itself and the target satellite.

To date, the Russian tests of the antisatellite system (or ASAT, in aerospace parlance) have been limited to low orbits and to Soviet test targets. But there is no discernible reason why ASAT's could not be boosted to higher orbits, to knock out satellites in geosynchronous positions. It would be comparatively simple for the Russians to destroy every early-warning satellite we have in orbit. Or all our communications satellites. Or to selectively destroy only a few satellites, as a warning to us. We could be blinded in space, or have our national communications network badly disrupted. By the end of this decade, when more than half our telephone and television communications will depend on satellite links, the US could be paralyzed by a satellite "blitz." From Wall Street to the Weather Service, satellites relay the information on which our nation runs. Destroy our satellite network and you cripple the United States.

The Department of Defense is working on ways to "harden" our military satellites, to protect them against intruders and destruction. These plans are classified, naturally, but they most likely include a combination of techniques, such as: (1) strengthening the satellites' structure to protect the electronics payload from damage; (2) making the satellites harder to locate, both by lowering their visibility to radar (*à la* the now-famous "stealth" aircraft) and by placing them in higher orbits; and (3) orbiting decoy satellites, so that the enemy won't know which satellites are the real targets and which are merely silvered balloons.

Of course, DOD is scrambling to develop its own ASAT system, on the theory that no one will destroy our satellites if they know we can retaliate in kind.

But our civilian satellites are neither hardened nor hidden. They are the most vulnerable targets, and the most valuable, in many ways. Would the Congress declare war against an enemy who knocked off our communications satellites? Perhaps an even bigger question is, Could we get the Congress together for a vote if our telephone system has been crippled by a satellite blitz?

There has been a low-key but consistent effort to produce a body of international law that will prevent armed conflict in space. In 1967 the United States, the Soviet Union, and many other nations signed and ratified The Treaty on Principles Governing the Activities of States in the Exploration and Use of Outer Space, Including the Moon and Other Celestial Bodies. The 1967 Outer Space Treaty (as it is mercifully abbreviated) prohibits the militarization of the Moon and the placing of "weapons of mass destruction" in space.

Fine. As far as it goes.

A ratified treaty has the force of federal law within the US. To break a treaty would be equivalent to violating any federal statute, such as the kidnaping law. The Soviet Union also takes its treaty obligations seriously. But while most Americans felt that the 1967 Outer Space Treaty ended the threat of military operations in space, the Soviets immediately showed that they had read the treaty's fine print very carefully, indeed, and had laid their plans accordingly.

"Weapons of mass destruction" is generally interpreted to mean nuclear weapons. The treaty forbids placing nuclear weapons in space, but it does not forbid ICBM's, which are based on the ground, or submarine-launched ballistic missiles (SLBM's). Although these missiles spend most of their 30-minute flight time in space, they are not considered space-based weapons.

The purpose behind the treaty's prohibition of "mass destruction" weaponry in space was to prevent nations from placing nuclear bombs in orbit, where they can hang overhead like a radioactive Sword of Damocles. While it takes an ICBM some 30 minutes to reach its target, an orbiting bomb can be de-orbited suddenly, reaching its target in five minutes or less after the button is pushed. ICBM's and SLBM's give the defending nation enough warning time to launch a retaliatory strike against the attacker. No time for anything more, except a few prayers or curses. But it is this threat of retaliation that prevents nuclear-armed nations from striking. Orbiting bombs cut that warning time down to virtually zero. If an attacker had sufficient warheads in orbit, it could demolish its victim before the defending nation had a chance to launch a counter-strike.

Hence the Outer Space Treaty's prohibition against "weapons of mass destruction" in space.

The ink was barely dry on the 1967 Treaty when the Soviets began testing their fractional orbit ballistic system (FOBS). This is a launching system that almost establishes itself in orbit, but not quite. It lofts a warhead almost all the way around the world, to deposit it on target without quite completing one orbit of the Earth. FOBS is within the letter of the 1967 treaty; yet it still confounds American plans for missile defense. Because FOBS can be launched southward from the Soviet Union, fly down the Indian Ocean, cross Antarctica, and strike at the US from the south — while all our defenses are looking northward, prepared against conventional missile attack from across the North Pole. FOBS puts us in the same position as the British were in Singapore in 1941, when their magnificent defenses bristled with guns pointing out to sea, while the Japanese attacked overland and took them from the rear.

There is nothing illegal about FOBS. But it, and the later ASAT system, have shown that the Russians have no intention of allowing outer space to remain a haven of peace.

Within the US, the main effect of the 1967 Outer Space Treaty was to shut down the only chance we had to build a spacecraft that could have reached out beyond the Solar System and headed toward the stars. Physicists Ted Taylor and Freeman Dyson had conceived a propulsion system that used compact nuclear explosives to literally push a spacecraft up to the tremendous velocity needed to fly beyond the Solar System. But the 1967 treaty prohibited placing "weapons of mass destruction" in space. No matter how Taylor and Dyson argued that their explosives were for propulsion, not destruction, the decision was to close down the program, which they had dubbed Project Orion.

However, even as the 1967 treaty was being ratified by the

nations of the world, technological developments were taking place that would make the treaty obsolete. In Massachusetts, at the Avco Everett Research Laboratory, the gasdynamic laser was being perfected. A new concept that combined a wind tunnel–like flowing gas system with laser physics, the gasdynamic laser made it possible to develop lasers of extremely high power.

I was the manager of marketing for Avco Everett at that time, and I recall the stunned silence in the Pentagon conference room when our physicists finished explaining the principles of the new laser. The James Bond film *Goldfinger* was playing its first run then, and here was an actual laser system that made Goldfinger's fictional weapon look puny. The work was immediately classified.

Within a few months, though, it became obvious that the Russians were on the same trail. The Soviets did not censor theoretical physics papers at that time. We could read their technical journals and watch them take step after step, leading to the point where they would realize the potential they were creating. Physics is the same throughout the universe; we cannot keep the workings of molecules a secret. Sure enough, just at the point where our physicists had realized that a gasdynamic laser was possible, the theoretical papers stopped appearing in the Russian journals. Their work became classified, too.

Avco Everett now sells "low-powered" gasdynamic lasers to the commercial metalworking market. The first of these was delivered to the headquarters of Caterpillar Tractor Company in 1972, where it was used to test new ideas of cutting the toughened steels that Caterpillar uses for building bulldozers.

The Avco Everett laser put out an infrared beam of 10 kilowatts (10,000 watts). Too low a power output to be classified, it is still more power in one minute than all the lasers built in the 1960s delivered, in total. That 10-kilowatt beam cuts through three-quarter-inch hardened steel at rates of 50 to 100 inches per minute.

That laser, which is rather primitive by today's standards, could be carried into orbit by the Space Shuttle.

The Soviets are now working not only on lasers of extremely high power outputs, but on particle beam weapons as well. Particle beam weapons emit streams of subatomic particles, such as protons. Lasers emit pure energy: light.

Many "experts" in the US and Western Europe have pooh-poohed the idea of beam weapons (particle or energy beams) in space as *Star Wars* fantasy. Yet much of the so-called futuristic technology in the *Star Wars* films is actually quite obsolete.

Beam weapons are not "weapons of mass destruction." They are weapons of pinpoint destruction. They are not capable of frying cities or burning wheatfields from orbit. But they can shoot down rocket boosters and destroy orbiting satellites.

Space is an ideal environment for beam weapons. There is no air to absorb the beam's energy or to "bubble up" because of the beam's heating effects and cause the beam to wobble and miss its target. In space, if you can see it you can hit it, with a beam weapon.

Picture the world of the 1990s. Resources are even scarcer than today; global population is half again as large, at least. Assuming we have not moved to develop the peaceful economic potential of space, the political tensions of today will be magnified tenfold in the 1990s. The US, the USSR, and perhaps other nations will be orbiting dozens, perhaps hundreds, of antiballistic missile (ABM) satellites, each armed with beam weapons. These satellites will be the first line of defense against missile attack.

The ABM satellites would lock onto an attacker's missiles while the rocket boosters are lifting them from the ground, which is the most vulnerable part of their flight, and destroy them with beams of subatomic particles or laser energy as soon as they cleared the upper reaches of the atmosphere. And it makes no difference if these boosters lift from a fixed missile silo, a submerged submarine, or from the elaborate racetrack network of an MX system. Once the boosters are rocketing through the air, they can be detected from space and destroyed. The MX system, all 30 billion dollars' worth, will be just as obsolete in the face of orbiting ABM weaponry as any other missile system.

Is this a step toward peace? A step away from nuclear holocaust?

Hardly. Because each nation knows it *must* be the first to complete a protective network of ABM satellites. The first nation to be fully protected can launch a first strike at its enemies and get away with it, because it can destroy the counter-strike long before the enemy missiles come close enough to be a threat.

We will want to be the first nation to achieve ABM protection. So will the Soviet Union. So will China and every other nation on Earth.

We could see, then, a new and deadly space race, where we launch as many ABM satellites as we can, as quickly as we can, and use them to destroy the other nations' ABM satellites as fast as they can put them up. And they, of course, will be doing the same to ours. A deadly, silent war, fought in secrecy a few hundred miles above our heads, without the public's slightest knowledge. A remotely controlled robot war, machines against machines, with no human casualties except for tension-induced illnesses among the controllers deep in their underground command centers.

The winner of this silent war will dominate the Earth. The loser will be open to nuclear devastation.

This space war could take place — may very well take place — within the next decade. Even sooner, the Space Shuttle will be carrying military astronauts into orbit for the more immediate missions of surveillance, communications, and command-and-control.

The Air Force plans to use the Shuttle in much the same way NASA and private users will: to make it cheaper and easier to place payloads in orbit. The Air Force is spending well over a billion dollars to erect its Shuttle launch complex at Vandenberg AFB in California. From Vandenberg, Shuttles carrying military satellites will be launched into polar orbits that crisscross the entire Earth twice each 24 hours. The Air Force is also developing the Inertial Upper Stage, a rocket propulsion system that can be attached to satellites to loft them into the high orbits that the Shuttle cannot reach. The Inertial Upper Stage will be made available for NASA and civilian satellites, as well as military payloads.

Today, military space operations are the province of automated, robot satellites. In the very near future, military astronauts will bring human capabilities to space: to launch bigger, more complex satellites; to check out and repair malfunctioning satellites; to return disabled satellites to Earth; possibly even to inspect or disable satellites launched by unfriendly nations.

Rumor has it that at least two of our military observation satellites have been blinded by laser beams fired up from the ground that ruined the optics systems aboard the satellites. In one case, the Pentagon admitted that a satellite had been blinded, but put the blame on a very bright forest fire raging in Siberia. This led cynics to the conclusion that the laser rumor was true.

As we saw in Chapter 20, the Solar Power Satellite has certain military implications. Its microwave beam might be used as a weapon, although the designers of SPS insist that such a possibility is out of the question. More likely, it will be necessary to defend SPS against terrorists or military threat: A ten-mile-long satellite makes an obvious target.

Solar Power Satellites could, of course, provide the electrical power for beam weapons in space. It seems much more likely that they would be used that way than as "death rays" to threaten civilian populations on the ground. In fact, if we reject the commercial development of space and forestall private construction of SPS, the military — not only of the US, but other nations as well — may very likely construct their own versions of Solar Power Satellites to provide the electrical power they need for their space-based weaponry.

The irony is that the more the Luddites work against private commercial development of space, the easier they make it for the military to totally dominate our space program. The best way to avoid an entirely military space presence is to move as strongly and swiftly as we can toward the *peaceful* uses of space.

Be aware that there will always be a military presence in space, as long as nation can lift sword against nation. But if we can begin to ease worldwide tensions by bringing the wealth of energy and raw materials from space to the peoples of the world, then we may be able to avert the military buildup that could lead to Armageddon.

Those who oppose a vigorous civilian space program are working toward a vigorous military space program, perhaps unwittingly, but the dupe is as effective as a dedicated soldier, in this case. We do not live in a peaceful world. The best way to avert war is to eliminate the causes for war. The best way to prevent the total militarization of space is to make a civil space program that is so productive that the military's only conceivable role will be to police the space lanes and protect our valuables from madmen and fools.

26

The Price

Beside, our losses have made us thrifty.
A thousand gildcrs! Come, take fifty!
—ROBERT BROWNING

In the final analysis, those who neither understand nor accept the space program plead poverty. We cannot afford to send billions of dollars rocketing off into space, they claim. I always get the feeling that they think the astronauts carry those billions along with them and, once on the Moon, lug huge pallets stacked high with greenbacks out of their landing module to leave them on the dusty lunar surface.

The money is not spent *in* space, it is spent *on* space. Actually, it is spent here on Earth, on jobs for American workers and profits for American companies. It is not merely the aerospace industry that makes profits. It is not scientists and engineers alone who receive paychecks. Every dollar spent on space has a multiplier effect in the national economy. Scientists buy groceries, believe it or not. And houses, clothes, cars, baby bottles, newspapers, hcating fucl, tobacco, liquor, furniture, household appliances, electricity, even toothpaste and toilet paper. They employ plumbers, housepainters, letter carriers, baby sitters, electricians, television servicemen, physicians, grocery clerks, secretaries, beauticians, barbers, dentists, and even psychiatrists.

The money is spent here. Over and over again. One dollar paid out to the space program changes hands at least eight times, according to the Chase Econometrics group, an economic analysis organization. This is the "multiplier effect" in the national economy, and space spending has one of the largest multipliers of any government program.

The money that leaves our country — and disappears from our economy — is the money we spend for foreign oil. The bitterest blow of our energy crisis is that it now costs us so much to buy foreign oil that we are becoming too poor to invest in the energy technologies that can replace petroleum. It's the old dream of the nineteenth-century robber barons come true: The petroleum producers have "cornered the market" on energy, forcing the price constantly upward (with our own thirsty demands for oil unabated) and simply carting off more and more of our national treasure in the process. In the seven years between 1973 and 1980 the cost of imported oil has risen more than 600 percent. In 1979 the US spent $59 billion on imported oil. That is nearly $5 billion per month; $6.7 million per *hour.*

OPEC is not the sole benefactor of this increased price of petroleum. The Energy Project of the Harvard Business School estimated that the OPEC price increases have raised the value of US reserves of oil and natural gas by at least $800 billion.

But the money that is spent on overseas oil is money that leaves the American economy. Five billion dollars a month in 1979; more than $8 billion per month in 1980. This is by far the biggest share of our annual $29-billion trade deficit. This monthly "oil tax" to keep petroleum flowing into our nation is what drives inflation, causes unemployment, weakens the value of the dollar, and — unless we change our ways — makes it so difficult to develop energy alternatives to petroleum.

As we have seen, space technology offers many ways of developing energy alternatives. Space technology also promises to open up the entire Solar System's treasury of natural resources, a bonanza of wealth so vast that no one can count it, as yet. But to reach these treasures we must spend billions on space. How much will it cost? Estimates for projects like the Solar Power Satellite or a permanent manned space station, are still very much in the "guesstimate" stage. Will it take $5 billion a year? Ten? Twenty?

Critics of the space program have pointed to the Apollo lunar landings and compared them to the great pyramids of ancient Egypt. "America's pyramids," they say. "A useless program initiated by the government to flatter itself."

I once asked one of these critics, "In what year did the pyramids break even?" He returned a blank stare. I explained that, although the pyramids must have cost a tremendous amount, they have also been bringing tourists into Egypt for more than four thousand years. Those tourists spend money in Egypt and, indeed, must be a major factor in the Egyptian economy. In what year, do you think, did the income from the pyramid-based tourist industry equal the original cost of the pyramids?

Neither he nor I nor anyone else knows. My suspicion is that the pyramids achieved fiscal breakeven sometime early in the days of the Roman Empire. I may well be wrong. Certainly they must be "in the black" by now. The point is, it does not really matter how much something costs, if it returns to you more than you spend acquiring it, and if that return comes soon enough for you to enjoy it. A space program that costs $20 billion per year, for example, will be cheap if it returns more than 20 billion dollars' worth of goods and services.

We now spend $20 billion approximately every ten weeks, for oil. Would the American people be willing to spend $20 billion *a year* on space? Even though that sum is small compared to our oil imports, it is still a large number of dollars, almost $10 a year for every man, woman, and child in the US. That's equivalent to asking the average family to forgo a night at the movies twice a year, or to give up a dinner at a good restaurant. Are the American people willing to make that kind of sacrifice?

Take a look at recent history, at the decade of the 1960s and the Apollo project. It cost the American taxpayer some $23 billion to reach the Moon, the equivalent of about three months of oil imports today. For that investment of $23 billion the US attained the technology and the trained team of people that have allowed us to go anywhere in the Solar System that we desire. By 1969 we had the ability to explore the Moon thoroughly, set up mining facilities there, build space stations in Earth-orbit, and even send expeditions out to Mars and the Asteroid Belt to seek the natural resources there.

Through much of that same decade we were also spending more than $30 billion a year on the war in Vietnam; an investment that returned us 50,000 dead, thousands of paraplegics, an Army broken in spirit, and a nation in tatters.

Also, through the decade of the 1960s, the federal government alone spent more than $500 billion on social programs, mainly under the "Great Society" banner which proclaimed a "War on Poverty" in the United States. The poor are still with us. In fact, according to official census figures, there are more poor in the US today than there were in 1960, and the gap between the rich and the poor is wider.

Throughout the 1970s, while the space program was being strangled by budget cuts, lack of goals, and slowdowns, more billions were poured out by federal, state, and local governments into social, welfare, educational, and employment programs.

The result? In 1980 unemployment hovered around the 10 percent mark nationally and over 15 percent for black men. Riots erupted in Miami and elsewhere. The major automobile manufacturers, once bulwarks of the national economy, reported losses in the billions of dollars.

Meanwhile, through the 1970s, one of the few bright spots in the nation's economic picture was the electronics industry. Computers, microprocessors, pocket calculators, and electronic games found worldwide markets that brought billions of dollars in sales to their manufacturers. Where did this technology for the home computer and the "microchip revolution" come from? Not from the Great Society bureaucracies.

Whenever I recite this economics litany in public, I am always asked, "What's in it for me?" Usually the questioner is a black man who sees that the space program may be great for white engineers who live in nice suburban developments, but has a suspicion that it won't do anything for the black ghettos of our cities.

Yet an expanding space program that has a positive impact on the national economy can improve life in the cities, even in the urban ghettos. When the economy improves, everybody benefits. Job opportunities increase at all levels. We have already seen that spending on space has a powerful economic multiplier effect: A dollar invested in space technology is spent over and again in our national economy, on goods and services as widely varied as titanium sheet metal and aluminum sandwich wrap, consulting engineers and supermarket checkout clerks.

One of the effects of the economic turndown of the 1970s was to slow the advancement of those who are on the bottom rung of the economic ladder. For the lowest-income groups to advance, we need an expanding economy, an economy that is demanding more workers, more goods, more services. The locked-in society that may result from the shutdown of our economic expansion will prevent the lowest-income groups from advancing. A vigorous space program that expands our economic horizons is the best chance for advancement that the poor have.

But more than statistics and economic policy is involved in this issue. We are talking about lives, about human beings who should have the opportunity to contribute to our society with all the heart and skill in them. That is why the story of Camden High School is so important.

RCA Corporation is headquartered in the Camden, New Jersey, area. When NASA first announced its "Getaway Special" program for the Space Shuttle, where an organization could buy five cubic feet of space aboard an early Shuttle flight for some scientific experiment, the RCA management decided to procure a Getaway space for Camden High School.

It was RCA's way of being a good neighbor. It was also good public relations for the corporation and might even encourage Camden youngsters to aim for careers in electronics. The RCA people knew that Camden High was one of the poorest schools in the state, smack in the middle of a crumbling urban ghetto. But, they reasoned, with the help of some company scientists and engineers, the kids ought to be able to produce a decent experiment.

What they found, once they actually visited the school, shocked them to their souls. Illiteracy. No books. Hostility to any stranger. Total indifference to learning. And yet, a spark was there, a yearning for something better than the ghetto had to offer. With lumps in their throats, the sleek white middle-aged RCA scientists and engineers "got down" with the black and Hispanic teen-agers of Camden High. Both sides learned a lot.

Beneath their studied mistrust of Whitey, the school kids were intrigued by the idea of producing something that would go into space. Instead of one science class being involved, the whole school got into "Orbit 81," as they soon named their project. A second ghetto school, Woodrow Wilson High, joined the project. The schools were transformed by excitement over their very own space program. Kids who could barely read began to put out a school newspaper — and learned reading, because they wanted to.

It was neither skin color nor economic level nor inherent IQ that held those kids in the ghetto. It was isolation from the real world, the world of opportunity. RCA presented the teen-agers with an opportunity to reach beyond the ghetto. The students responded.

Their experiment — to study an ant colony under zero-gravity conditions — will fly in an early Shuttle mission. Some of those students will go on to college and careers; others, no doubt, will remain in the ghetto for life. But the excitement and the opportunity of space will have changed at least a few ghetto lives for the better.

Using space as a spur, other corporations can rescue other teen-agers from their ghettos. What is the economic value of that? To get back to more formal economic data, consider the study published in 1976 by Chase Econometrics. It was received with yawning indifference by the media, and most taxpayers don't even know it exists. But the Chase study showed that for every billion dollars invested in the space program on a sustained basis between 1975 and 1984, the effects on the economy would include:

1. An increase in the gross national product of $23 billion.
2. Creation of more than 800,000 new jobs.
3. In terms of "rate of return" (which would be called "profit" in a private firm), each dollar spent would produce a 43 percent return.
4. The "multiplier effect" for dollars spent on space is somewhere between three and eight; this means that every dollar spent on space has the effect of three to eight dollars' worth of new purchasing power.

Consider also the value of the aerospace industry to our balance of international trade. In 1979, aerospace exports (which includes sales of commercial airliners) totaled nearly $13 billion, second only to agricultural exports, which amounted to nearly $30 billion. But our agriculture exports must be weighed against the agricultural products that we import each year, such as sugar, rice, cocoa, coffee, and so on. This amounted to more than 50 percent of our exports in 1979, while the aerospace trade balance is almost 90 percent in our favor.

Space technology pays its own way today and returns us sizable profits, even now. NASA's annual budget over the past few years has run from $4 to $5 billion per year. That seems like a huge amount of money, at first glance. But compared to what we routinely spend elsewhere, the NASA budget is trivial. Compared to the riches that we could reach in space, that budget is ludicrously, perversely, small.

The Department of Health and Human Services (formerly Health, Education, and Welfare), which handles most of the federal health and welfare funding, receives more than $200 billion annually. This means that HHS spends the equivalent of the entire NASA budget every ten days. Five billion dollars spent every ten days, month after month.

The Department of Defense, with a budget of some $150 billion per year (and rising), spends NASA's paltry $5 billion in a couple of weeks. Every couple of weeks. In fact, the DOD budget for research and development is larger than the entire NASA budget.

No one wants the US to be weak militarily or to ignore the needs of the poor and the sick. We should keep things in perspective, though. How much is enough? How high is up?

It isn't only the government that spends billions of dollars. In 1978, Americans spent as much money on McDonald's hamburgers as we did on space; one and a half times more on pizza than on NASA; three times more on cosmetics; twice as much in A&P supermarkets; one and a half times as much for toys. We spent $5 billion in 1978 for potted plants, seeds, and flowers; $7.4 billion on spectator amusements; $17.9 billion on cigarettes and tobacco products; $19.5 billion on radios, television sets, audio records, and musical instruments; and $30.9 billion on alcoholic beverages.

Nor is NASA's budget extravagant by the standards of corporate business: IBM's *profits* were three-quarters of the NASA budget in 1978. In January 1980 Mobil Oil felt obliged to take out display ads in the *New York Times* and other major newspapers to explain that, yes, $2 billion in profits seems like a lot, but that's not what makes gasoline so expensive. While pointing out that building one North Sea drilling platform can cost $2 billion, and a new oil refinery costs more than $1 billion, Mobil neglected to mention that its 1979 sales totaled $79.1 billion and the $2 billion in profit they were explaining away was quarterly.

Look at the NASA budget in comparison to what we spend on other modes of transportation: The federal government alone, in 1979, put out more than $11 billion for various ground transportation projects. More than $6 billion went into highway construction and maintenance. And this does not include the billions more spent by state and local governments. More than $2 billion was spent on highway safety in 1979; yet we still manage to kill some 50,000 people per year in highway accidents. Airport construction and equipment took nearly $3 billion in federal funds in 1979. Again, state and local governments spent about an equal amount for airports.

Will a vigorous space program cost us as much as we spend on cigarettes each year? Or as much as drunk drivers cost us ($6.5 billion in property damage annually)? One thing is certain: The returns we get from space will be more valuable to us than the results of cigarette smoking, or drunk driving, or the hundreds of billions we have squandered to build bureaucracies to deal with the poor.

The money we spend on space will yield us the Solar System: energy, raw materials, a new frontier, jobs and profits here on Earth, a stronger nation, a healthier economy that will do more for the poor than any welfare program.

Can we afford to move boldly into space? Can we afford not to?

27

The New Politics

When NASA announced that the first Space Shuttle would be named *Constitution*, the agency was deluged with thousands of protest letters demanding that the ship be named *Enterprise*, after the starship in the television series *Star Trek*.

The first step in the new politics of space had been taken.

As we have already seen, science fiction writers and readers helped to create a climate of opinion favorable to space flight, back in the 1940s and '50s. When Sputnik streaked into the October sky in 1957, Americans were not so much startled that someone could place an artificial satellite into orbit as they were shocked that it was the Russians who had done it. They had tacitly assumed that Americans would be the first into space. It was that way in all the stories, wasn't it?

I know of only one science fiction story that foretold of the Russians getting into space ahead of the Americans, and that one was never published. It was a novel of mine, my first novel, and like most first novels it is probably a blessing that it was never put before the public. Written in 1949 and 1950, its plot was based on the idea that the Soviet Union would be the first to get into space, but the Americans would launch a highly secret, Manhattan Project–type of program that would place an astronaut on the Moon before the Russians could get there.

Publisher after publisher rejected the novel, with the usual form letters. Finally one editor was kind enough to invite me to his office and explain that his company would not publish the novel, even though it was "no worse than most of the science fiction we print." The main reason for their rejecting the novel was that Senator Joseph McCarthy, then at the height of his power as a witch-hunter, would persecute any writer, editor, or publisher who dared suggest that the Russians were in any way superior to Americans.

Of course, today that novel is history, not science fiction.

When the United States did begin to move vigorously into space, the science fiction writers were pushed aside by the engineers and administrators and astronauts. Some of the writers — especially leftish writers like Britain's Brian Aldiss — joined the naysaying chorus that portrayed Apollo as a "Moondoggle."

Then came the backlash that virtually wiped out the American space program. Most science fiction writers stood mute, either agreeing that Earthbound problems like pollution and government corruption were more important, or stunned that such folly could be true.

Arthur C. Clarke, ten years after the Apollo 11 flight, wrote:

When Neil Armstrong stepped out onto the Sea of Tranquility ... no one had ever dreamed that the first chapter of lunar exploration would end after only a dozen men had walked upon the Moon ...

But in the 1970s a strange phenomenon began to take place in the United States and throughout the Western world. Space flight virtually disappeared from the news. Although the Russian program continued unabated, it was barely noticed in the Western media. There were no more American astronauts going into orbit, no big news splashes except for occasional probes of Mars or other planets. Launches of communications satellites or weather satellites were so routine that no one took any notice of them.

The biggest news story of the 1970s, as far as space was concerned, was the fiery death plunge of Skylab.

Yet, while the media and the politicians turned their backs on space efforts, interest in science fiction was spreading rapidly throughout the industrialized world. Sales of science fiction books boomed. Students organized science fiction classes on college campuses, in high schools, and even in junior highs. The academic world, sensing a new ecological niche (complete with tenure) responded by creating science fiction curricula — often taught by members of the English department who knew less about the subject than their students.

While interest in space declined, interest in science fiction and futurism advanced. Science fiction crept into commercial television and then into motion pictures. Often jejune dramatically, it was nevertheless brightly, brassily, optimistic. Old-fashioned virtue always triumphed over totalitarian evil. The best example of this phenomenon is Gene Roddenberry's *Star Trek* television series. It showed an intriguing future world of stalwart heroes, simple virtues, and glittering future technology. Seemingly sophisticated men and women, including millions of college students, stopped everything to watch *Star Trek* each week — and later, daily.

When NBC tried to kill the series, tens of thousands of letters poured in from angry "Trekkies." When the show was finally dropped from prime time, reruns popped up in almost every TV station in the nation, to be played daily. The Trekkies watched so faithfully that they could lip-sync the dialogue. *Star Trek* became a cult, almost a religion. Its audience today is far larger than it was during its prime-time era.

So when the first Space Shuttle was to be named, it was no surprise to the science fiction fans that the Trekkies rushed to their typewriters and forced NASA to dub the craft *Enterprise.*

The second step in the new politics of space involved the Viking spacecraft, on Mars. The Viking craft that reached Mars in 1976 — two orbiters and two landers — looked for evidence of life on the Red Planet. The results of their automated investigations showed no conclusive evidence for the existence of Martian life. The media took this to mean that there is no life on Mars. That negative interpretation may turn out to be correct, but the question is not quite closed. Not yet. Carl Sagan, the driving soul behind the Viking program, has argued forcefully that Martian life must be very different from our own, and that the Viking results show just what one might expect from a different kind of biology.

But most scientists refuse to climb out on Sagan's limb; they stick to the safely conservative position that no evidence for life has been found on Mars.

"No life" equals "no interest." By 1979, officials in the Viking program office at NASA were beginning to wonder how long they would be allowed to keep the Viking instruments working. The spacecraft had originally been designed to operate for a minimum of nine months after reaching Mars. Four years later, they were still "on the air."

It costs money to keep a big radio antenna tuned to Viking's signals, and more money to keep technicians at the task of reducing those signals to meaningful data that can be turned over to research scientists studying Mars. With constant cutbacks in funding, NASA administrators began to ask how long they could, or should, keep Viking "alive."

Years earlier, faced with a similar funding cutback, NASA had quietly turned off all the automated instruments left on the Moon by our Apollo astronauts, even though they were still functioning and transmitting information about conditions on, around, and under the lunar surface.

By the end of 1979, the Viking I lander and orbiter were both still functioning, although the Viking II pair had shut down. Early in 1980, the Viking I orbiter expired. But there was still the sturdy little Viking I lander, photographing the Martian landscape, sniffing the tenuous air, and sending back weather reports daily. Since its touchdown on July 20, 1976, Viking I had shown Earthbound scientists a panorama of Martian seasons, complete with photographs of frosts forming on the rocky Plain of Chryse.

It had cost slightly more than a billion dollars to get the Viking spacecraft to Mars. Now NASA was seriously contemplating turning off the spacecraft's instruments, for lack of a few hundred thousand dollars. A handful of private citizens decided to do something about it. They created the Viking Fund, a private charitable organization dedicated to raising enough money to keep the Viking I lander alive and functioning as long as its nuclear power source would allow.

The Viking Fund is the creation of two Englishmen: science writer Eric Burgess and astronautical engineer Stan Kent. Burgess has been writing about rocketry and astronautics since the 1950s. Living in California, he became chairman of the Sacramento chapter of the American Astronautical Society and launched the idea for public donations to a Viking Fund from there. The main spokesman for the Fund has been Kent, an energetic young man who looks and sounds more like a rock singer than an astronautical engineer. Kent left his native England to join the American space program, and quickly became active in the Viking Fund.

The goal of the Fund is both simple and subtle: to raise enough money to keep the Viking program alive. At first the enthusiasts aimed to raise a million dollars. Using strictly volunteer help and scrambling for publicity wherever they could find it, the Viking Fund people soon realized that it would be impossible to meet such an ambitious goal. But in January 1981 they handed a check for $60,000 to officials of NASA, in a ceremony at the Smithsonian Institution's Air and Space Museum, in Washington, D.C.

NASA was at first very cool to the idea of accepting private donations. Although the NASA charter legally permits such action, it also specifically states that NASA cannot obligate itself to spend donated money for any specific purpose. In other words, NASA could take the Fund's money, but was not about to promise to use that money as the donors wanted. However, after months of delicate negotiations, NASA's upper echelons saw the wisdom of accepting the money and putting it to work directly on Viking.

Down in the trenches, in the Viking program offices in Washington and the Jet Propulsion Laboratory, there was rejoicing.

The subtle part of the Viking Fund is the politics. Burgess, Kent, *et al.* wanted to show Washington that there is public support for the space program that goes far beyond the politicians' understanding. As Kent put it, a million citizens donating one dollar each would have a much stronger impact on Washington than one donor putting up a million dollars.

The specific, goal-oriented nature of the Viking Fund serves to focus space-minded citizens on that goal and gives them a specific action that they can easily understand and accomplish. The Fund is an important step toward creating an electorate, a pressure group, a grassroots movement in favor of a stronger, bolder space program.

This is the way the new politics of space is being shaped. All across the United States, and in Europe as well, grassroots organizations are springing up. In America, some of these groups made an abortive effort to involve themselves in the Presidential campaign of John Anderson, the third-party candidate in 1980. Although Anderson was soundly defeated, many of his supporters continued to work with the grassroots space activists and to help them form political action committees (PAC's).

In this effort they are following the lead of the Trekkies, who in turn patterned their efforts after the earlier science fiction fan clubs. Science fiction fans have organized themselves into clubs since the 1930s, and regularly hold conventions in almost every major city in the United States. "World" conventions have been held since 1939, and "WorldCons" have taken place in London, Heidelberg, Melbourne, and elsewhere.

The grassroots organizations that favor stronger efforts in

space are not modeling themselves merely on the science fiction and *Star Trek* clubs. They are also borrowing heavily from the environmental movement of the 1960s and '70s. This is because the space movement has as its goal *political action.* Like the environmental movement, the space movement is rising from the grassroots of the American political system, crossing party lines and ideological affiliations: Liberals and conservatives, rich and poor, young and old, engineers and Trekkies are banding together to build a political force that will move Washington toward a new, vigorous space program aimed at solving economic and social problems here on Earth.

In a few places the environmentalists have joined their space brethren. For the most part, however, the dedicated environmentalists see a strong space program as counter to their own goals. The environmentalists are largely aligned with the decentralist views of the Luddites. Since a big space program means Big Government and Big Business — the Establishment in all its centralized power — many environmentalists are wary of the space movement.

What is needed is a melding of the two interests — and strengths — into a unified program of action. Space proponents point out that a strong space program will help to protect Earth's environment: Extraterrestrial resources could put an end to the gutting of this planet for metals and minerals; orbital factories could move prime sources of pollution off the Earth altogether; new technology could revolutionize transportation and living styles. The long-range goals of both the space and environmental movements are similar: a clean, green Earth, a safe and fitting home for humankind. If the two movements can be merged, if the individuals who comprise these movements can see their common interests as more important than the areas in which they conflict, then the entire human race may move a giant stride closer to the healthy kind of world we all want to live in.

This may be the most important political movement in America today, the contact between the space enthusiasts and the environmentalists. Together they can lead the way to a second-generation technology, not merely in space but here on Earth as well, that could achieve their common goals.

It is clear that to build a new space program, we must change the order of business in American politics. The veterans of the earlier space program are leaving the arena, either through retirement or death. Von Braun, Russia's "master designer" Korolev, the inspirational pioneers such as Tsiolkovsky, Goddard, and Oberth have all died. So have the political leaders who created the space race of the 1960s: John Kennedy and Nikita Khrushchev. The earliest astronauts and cosmonauts are mostly out of the space program now, and the vast teams of engineers and scientists who worked with them have either left the space program or retired altogether.

As Kennedy said in his inaugural address, "The torch has been passed to a new generation of Americans." Whether we want this responsibility or not, it is *this* generation of men and women who will either create the political environment for a vigorous space program, or see the United States — and the world — spiral down into decay and disaster. This generation is at work, at the grassroots level. The movement is gaining strength. It spans the nation and it is as diverse as any melting pot's brew can be. The grassroots movement for space includes at least a hundred known organizations, plus hundreds of other local clubs and interest groups that have yet to be identified in any formal listing. These groups range from professional engineering organizations, like the American Institute of Aeronautics and Astronautics (AIAA), to *Star Wars* fan clubs. There are citizen-support groups in the fields of education, research, fund raising, and political action. Together, they comprise at least a quarter-million members, by the most conservative estimates. On a national basis, this is a small number, only a tiny fraction of the millions of Americans who have formally joined environmental groups, for example. But the space movement is very new; it began to organize only in the last years of the 1970s. And it is rapidly growing.

Among the organizations in the space movement are:

• The L-5 Society, whose primary aim is to build huge permanent colonies in space that eventually will house millions of people. To reach this goal, the L-5 Society sponsors educational meetings, publishes a newsletter, and even supports a lobbyist in Washington. (For further discussion of L-5 colonies, see Chapter 28.)

• The Universities Space Research Association, a consortium of more than 50 universities and as respectable as a church, which operates the Lunar and Planetary Institute and the Institute for Computer Applications in Science and Engineering for NASA.

• A whole alphabet soup of professional scientific and engineering organizations, such as AAS, AIAA, the Aerospace Industries Association (AIA), the Aerospace and Electronics Systems Society of the Institute of Electrical and Electronics Engineers (IEEE), and many more.

• The National Space Institute, a nonprofit corporation that includes among its members television newsman Hugh Downs, singer John Denver, and some 23,000 other private citizens ranging from science fiction writers to corporation presidents.

• The Space Sciences Institute, a private research organization established by Princeton University physicist Gerard K. O'Neill (who originated the L-5 space colony concept) to fund space-related research that is not funded by NASA or other government agencies.

• The Planetary Society, founded by astronomers Carl Sagan and Bruce Murray to create interest and support in exploring the planets of the Solar System — and perhaps beyond, eventually.

• Science fiction clubs all over the world must be included in this list, as well as the still-formidable organizations of *Star Trek* fans, several political action committees, plus organizations such as The Futurian Alliance, the Citizens for Space Demilitarization, the Organization for the Advancement of Space Industrialization and Settlement (OASIS), and so many others that it is difficult to keep track of them all.

It's a polyglot group, certainly, with almost as many different opinions among them as there are individuals. But on one point they are agreed: The United States must move into space and utilize the resources there to solve our problems at home and abroad.

28

Colonies in Space

> More important than the material issues
> ... the opening of a new, high frontier will
> challenge the best that is in us ... the new
> lands waiting to be built in space will give us
> new freedom to search for better govern-
> ments, social systems, and ways of life ...
> —GERARD K. O'NEILL

No discussion of the capabilities of space technology would be complete without an examination of the concept of space colonization popularized by Professor Gerard K. O'Neill, and the L-5 Society. This kind of space colony was not predicted by science fiction; it was literally invented by one of O'Neill's Princeton classes.

Although science fiction has been intimately connected with the space program since before its beginning, there are at least two embarrassing failures of science fiction to predict major events in space technology: The O'Neill-type space colony is one. The other concerns the first flight to the Moon.

Stories have been written about flights to the Moon since Plato's time. Johann Kepler, the great German astronomer, wrote "Somnium," a fantasy published after his death in 1630, in which a man travels to the Moon in a dream. Cyrano de Bergerac wrote his "Voyage to the Moon" a generation later, and even had his hero use rockets. Jules Verne, H. G. Wells, Robert A. Heinlein — every major and most minor science fiction writers took a crack at the first-flight-to-the-Moon story. Not one of them predicted that the first lunar landing would be televised back to billions of eager watchers on Earth.

Similarly, the concept of building mammoth colonies in space, where nothing but emptiness exists today, never originated in a science fiction story. Science fiction writers (myself included) always assumed that the first permanent human settlements would be on or beneath the surface of another world: the Moon, Mars, Venus, for example. Plenty of stories were written about space stations in orbit around the Earth. George O. Smith wrote a famous series about a station placed along the orbit of Venus. But they were always seen as way-stations, outposts, not permanent colonies.

"Planetary chauvinism" is the way a rueful Isaac Asimov described the writers' failure to imagine colonies built in empty space.

In the autumn of 1969, two months after the Apollo 11 landing at the Sea of Tranquility and at the height of Vietnam unrest on campus, O'Neill was assigned to teach a freshman course in physics. In an attempt to get the students more interested in the work, he prepared a list of informal challenges for them to ponder. The first one was, "Is a planetary surface the right place for an expanding technological civilization?"

He never got to ask the next question.

The students took off with that problem and literally invented what has come to be called the L-5 space colony concept: gigantic space habitats built between the Earth and the Moon, big enough to house thousands or even millions of permanent residents in a completely Earthlike environment. The concept caught the imaginations of people everywhere and received enormous publicity in media as diverse as *Physics Today,* the journal of the American Institute of Physics, *Time* magazine, and the *60 Minutes* television show. It even caught the attention of Senator William Proxmire, a notorious Luddite, who promptly snapped, "Not one cent for this nutty fantasy."

Because it is such a grandiose scheme, the L-5 space colony concept has spawned almost as many enemies for space technology as friends. The uninitiated see drawings of a fantastic, ten-mile-long colony floating in space, surrounded by solar mirrors and space factories, and immediately conclude that this is "far-out science fiction stuff," or, as Proxmire put it, "nutty fantasy."

On the other hand, L-5 enthusiasts have a religious fervor in them. They see space colonies as the answer to all the problems of the human race, from overpopulation to environmental degradation, from threats to individual freedom to threats to endangered species.

To me, the idea of building these giant space habitats *before* space industries begin earning profits is putting the cart far before the horse. Space colonies will come about, I am certain, but only after earlier space endeavors have generated the capital to build such colonies.

The basic idea behind the L-5 concept is elegantly simple: By building colonies in empty space, it is possible to create habitats that are totally Earthlike. A colony on the Moon (or, more likely, beneath the Moon's surface) would exist in the Moon's one-sixth gravity. There is no way around that. It would also tend to be small and cramped if it has to be dug out of the lunar rock.

Colonies on other worlds, such as Mars, Venus, Mercury, and the moons of Jupiter or Saturn, would have to dig themselves into the ground or take their chances with the surface conditions of that world. Each would be in a hostile environment, faced with low gravity. Worse still, such colonies would invariably upset the natural environment of the worlds on which they are planted.

The L-5 concept neatly sidesteps such problems. The colony is built in empty space, where there is no "native" environment to despoil. The interior of a large-enough habitat can be made to resemble Earth's environment almost perfectly. Not only that, but a space colony of ten miles' size can have an Earthlike feel of gravity inside it, and still have virtually zero gravity outside!

As O'Neill has put it, why climb out of Earth's 4000-mile-deep gravity well merely to sit yourself down at the bottom of some other world's gravity well? In a space colony the gravity well could be virtually nil, allowing cheap and easy transport of people and materials to and from the colony.

The term *L-5* goes back to the work of Comte Joseph Louis Lagrange, the Italian-French astronomer and mathematician of the eighteenth century. He found that there are certain points between any two astronomical bodies that are gravitationally stable. That is, an object placed at one of those points would tend to remain there, rather than drift away. There are five such areas between the Earth and the Moon. Two of them, called L-4 and L-5, lie along the orbit of the Moon, sixty degrees on either side of the Moon's position. The L-4 and L-5 regions can be thought of as two invisible harbors that travel around the Earth just as the Moon does, always remaining precisely equidistant from both the Earth and the Moon: a quarter-million miles, roughly, from where you are now.

Place an object in the L-5 region and it will stay there indefinitely, orbiting around the Earth just as the Moon does. Satellites in other orbits between the Earth and the Moon tend to drift out of position over time, but objects at L-4 and L-5 will stay put.

A space colony could be built almost entirely from ores scooped out of the Moon's surface. Materials that the Moon lacks, such as nitrogen and water, would have to be hauled up from Earth or found elsewhere. O'Neill has shown that a colony big enough to house 10,000 residents would need no more structural metal than a modern ocean-going supertanker.

Once the space colony idea began to take form among the Princeton students, O'Neill saw that his earlier work on electrically driven particle accelerators could be turned into a "mass driver" that could catapult payloads of ore off the Moon to a "mass catcher" waiting at the L-5 position or elsewhere in space.

Many different designs for space colonies have been studied. The earliest were huge cylinders, as long as Manhattan Island and four miles in diameter. Other designs look like doughnuts and spheres, and one is even dubbed "Sunflower."

The interiors of these colonies, whether they be cylinders, toruses, or some other shape, are landscaped with hills and valleys, woods and streams, small lakes and little towns and villages dotted among the grassy grounds. Farms and factories are placed in separate structures, outside the main habitat, where each individual module can maintain its own special environment. A farm for tropical fruits, for example, would be kept at a higher temperature and humidity and a different day-night cycle than a laboratory for molecular biology. Factory modules would be studded with focusing mirrors for creating high-temperature heat, while nearby there might be solar parasols to create the shadow needed for low-temperature cryogenic work.

The whole colony would be solar powered; it will be in sunlight constantly. Huge mirrors would capture solar energy, and would be programmed to provide the desired day-night cycle within the colony's main habitat area.

Inside the main habitat, an Earthlike one-gee force would be maintained by spinning the entire structure at the proper rate. Small rockets on the outer shell would start the colony spinning; in the frictionless vacuum of space that should suffice for a period of indefinite length, except for minor corrections now and then.

While the colonists would feel a completely normal "gravity" inside the habitat, just outside the structure there would be zero-gee landing docks for spacecraft. And the colony's exports of manufactured goods, foods, medicines, whatever, could be launched Earthward or Moonward with the barest of nudges.

The colony will be bathed in harsh radiation from the Sun and cosmic rays, so it will need a protective coat of rocky rubble from the Moon's surface. From the outside the colony may look like a very large asteroid — but an asteroid peculiarly studded with antennas, mirrors, and other structures.

When the L-5 idea first hit the public, some people saw space colonies as a solution to population problems. Simply build lots of colonies and ship out the excess population! But without even thinking about the social, political, and ethical problems such an undertaking would entail, the sheer physical problems quickly scotched that idea. There isn't enough rocket propellant on Earth to lift enough people to put a dent in the population growth. Even if you could "remove" a million people a year, or a hundred million — which means launching nearly 275,000 people *per day* — you would barely stay even with the global population growth rate. Take a hundred million people from five or six billions and what do you have left? Five or six billions.

Here, for once, science fiction writers were ahead of the game. They had considered "exporting" population problems back in the 1930s and quickly realized that it would not work.

O'Neill and the L-5 enthusiasts soon understood that they needed some economic or political rationale to support their desire to build space colonies. They hit upon Glaser's idea of Solar Power Satellites. Space colonies will be necessary, they reasoned, as manufacturing centers for SPS's. However, their desire to build a colony first and then start manufacturing SPS's from lunar materials is a bit like wanting to build a luxurious housing development on Alaska's North Slope before drilling the first oil wells there.

L-5 is a dream, a good dream, one that is needed to keep the long-range possibilities of space technology before our eyes. Humans will live in space permanently, raise families there, and expand through the Solar System. The mining craft that ply the Asteroid Belt will be modified L-5 colonies, housing hundreds of families in terrestrial comfort and containing their own smelting plants, rather than space-going tugboats operated by grizzled loners.

But before anyone invests the hundreds of billions of dollars needed for an L-5 space colony, hundreds of billions in profits must be generated by space industries. Energy from SPS's, raw materials from the Moon and elsewhere, manufactured products from space factories — all these will precede space colonization, not follow it. Unless . . .

An enterprising publisher asked me in the early 1970s to write a novel about the very first space colony. I found my characters constructing a colony at the L-4 position (because it offers a prettier view of the Moon) for reasons that have little to do with Solar Power Satellites or mining the asteroids. In my novel, *Colony*, the first space colony is built by a consortium of multinational corporations because the heads of those corporations want a safe place to live in their accustomed luxury when the Earth goes through the collapse foreseen for the early twenty-first century. These corporate leaders do not realize, of course, that when that collapse begins, their big, luxurious "country club in the sky" will be a prime target for terrorists.

If and when space colonization becomes a reality, the fact that a few thousand human beings have made their homes in space will have an enormous impact on the lives of the billions who remain on Earth. The advent of human settlements in space will have far-reaching effects on the political, social, and economic lives of all humankind.

For the first time since 1776 human communities will have not only the opportunity, but the necessity, of creating new social rule books for themselves. Many of our old problems of social class, ethnic background, cultural expectations, education, and economic status will be quite meaningless in the spanking-new environment of a space colony. The colonists will, at the outset, be largely classless, relatively homogeneous, and highly motivated. As time passes and each colony evolves its own social system, conditions will naturally change. But at the start, the atmosphere will resemble that of a pioneer, frontier settlement: You are valued by your neighbors for your deeds, not your history.

Inevitably, space colonies will have an economic impact on Earth. We have seen that as manufacturing moves into space, Earth will become cleaner and greener. But what happens when space colonies take the lead in heavy industry and hold the economic life of Earth in their hands?

Historically, whenever a colony has become self-sufficient it has cut itself free from its motherland. This helped bring about the collapse of both the Roman and British Empires. On July 4 we celebrate the moment when Americans finally recognized that we could get along perfectly well without Britain's help.

The chances are excellent that as the citizens of space colonies gather in their tree-fringed town commons, one inevitable topic of conversation will be the prices and taxes imposed by those unfeeling money-grubbing flatlanders Back Home. Once the colonists realize that they are self-sufficient, that they can live without the markets and governments Back Home, that they have the whole Solar System at their fingertips — they will strike for independence. Once they realize that Earth is dependent on them for energy, for raw materials, for manufactured goods and medicines and even food — they will win their independence.

These are some prospects for the twenty-first century. We may not see them, although our children almost certainly will. Real though they are, they are not as immediate as the economic, political, and military possibilities we have examined in the preceding chapters.

Space colonization will be a reality. Our children will build one after another and use the colonies not merely to house themselves halfway between the Earth and the Moon but to sail outward to the asteroids and beyond, perhaps even on toward the stars in journeys that will take hundreds of generations to complete.

Such colonies and arks in space will transform the human race into a truly spacefaring species. No longer bound to the surface of one tiny world, our kind could ultimately become an immortal species, able to survive the death of its home planet or even the wreck of its entire Solar System.

Like gleaming pearls on an invisible linkage of radio waves, space colonies and arks will spread across the void, carrying the human race to its destiny among the stars, taking us, as Robert Frost said:

> *To the land vaguely realizing westward,*
> *But still unstoried, artless, unenhanced,*
> *Such as she was, such as she would become.*

IV

AGENDA FOR ACTION

We the People of the United States, in Order to form a more perfect Union, establish Justice, insure domestic Tranquility, provide for the common defence, promote the general Welfare, and secure the Blessings of Liberty to ourselves and our Posterity, do ordain and establish this Constitution for the United States of America.

29

Future Four

The year is A.D. 2000.

The world is far from at rest. Global population has passed six billions. The famine that began two years ago in Bangladesh has spread across India and into the tribal states of south-Saharan Africa. In the wake of starvation comes disease. Shipments of food, medicine, and trained medical personnel simply cannot keep up with the monumental size of the problems. Millions are dying.

The government of India is desperately rushing to complete its first rectenna farm for the Solar Power Satellite now orbiting over the Indian Ocean. Critics say that India is wasting money on a futile gesture, but the official government policy is to accept the offer of energy from space that the North Atlantic consortium has made and to use the energy to create new jobs and income for millions of unemployed Indians.

Population has kept stable in North America and Europe, where the industrialized nations have pooled their technological skills to bolster the economies that were tottering so badly less than 20 years ago.

Oil from the Middle East is still important to Europe and the US, but less so each year. Hydrogen fuels have largely replaced petroleum for ground transportation, and the Soviet Union is now exporting magnetohydrodynamic power generation equipment to nations that have coal to burn and need MHD's efficiency and environmental cleanness.

The first Solar Power Satellite is beaming five gigawatts of energy to its rectenna farm in New Mexico, and the US balance of trade has been in the black for the second year in a row, thanks mainly to the export of energy technology.

The permanent space stations in orbit around the Earth now number six, all of them open to UN inspection. The lunar mining consortium promises to start delivering next year the raw materials for building eight more SPS's. And the UN-sponsored Mars expedition is within a month of landing its thirty explorers on the surface of the Red Planet.

A US Presidential election is coming up in November. One of the main issues shaping up is the question of how quickly we can phase out the old-fashioned nuclear power stations, and where we should store their radioactive wastes, which have been accumulating on the Moon.

30

Insure Domestic Tranquility

> Nothing can survive on the planet unless it
> is a cooperative part of a larger, global life.
> Life itself learned that lesson on the primi-
> tive earth.
>
> —BARRY COMMONER

The space program cannot be separated from all the other activities and problems, dreams and fears of humankind. Just as our planet does not exist isolated from the rest of the Solar System, any space program we undertake is merely a part of the larger whole of human endeavors.

It is no coincidence that the American economy began its downward spiral at the same time we began to cut back on our activities in space. Nor is it coincidence that American prestige and political power have crumbled also during the same decade.

In *The Decline of U.S. Power* (Houghton Mifflin Company, 1980), Bruce Nussbaum, *Business Week*'s associate editor for International Money Management, wrote:

> ... the economic base of U.S. ... strength is as important a projection of general U.S. power around the world as the number of missiles the country has aimed at the Soviet Union.

Mao Tse-tung could write that "political power grows out of the barrel of a gun," but the real strength of a nation stems from its economic power. You must *buy* those guns. Even if you manufacture them in your own country, you must be able to pay for them. And for the food and all the other goods and services that make a nation strong.

A vigorous, well-directed space program can strengthen our floundering national economy. Not merely by creating additional jobs for workers or profits for industry, but by bringing real wealth—in the form of energy and raw materials—from space to Earth.

The history of our space effort shows that space technology builds new industrial muscle. Aerospace industry exports are one of the few bright spots in the US international balance of trade.

Because the technical challenges of space are so stringent, space technology is at the cutting edge of our scientific and engineering capabilities. Invariably, the new technologies created in the space program lead to new industrial capabilities on Earth. The microelectronics revolution that is sweeping the world with microprocessors, home computers, video games, and even robots, stems directly from space technology.

All around us we hear cries that our economy is going under, productivity has turned negative, American ingenuity has dried up. We hear demands for a "reindustrialization of America." But reindustrialization does not consist of aiding failing old industries like steel and automobiles. That is investing in the past. We must invest in the future.

"One way to get private business moving again is to increase spending on research and development," Nussbaum wrote in *The Decline of U.S. Power.* Decrying corporate managers who are "more concerned about the fast buck than about future investment . . . more interested in buying other companies or their own stock than in planning for the next century," he insists that the road to economic health lies in R&D investment that will create new sources of profits and jobs. "A rallying point is needed to draw together the forces of growth," he states, "and whether the goal that is set is the conquering of space, the oceans, or the cities themselves, it must be seen as an American goal."

Conquering space is a dubious concept: We can no more conquer space than we can turn ourselves into butterflies. But we can *use* space, use the environment and resources of space as a new source of wealth, a gold mine of energy and raw materials, a new frontier that can generate entire new industries and revitalize existing industries.

Without using the wealth in space, we cannot revitalize the economy. We cannot salvage the cities, or stem the inexorable tide of population growth and resource depletion. We can create a new industrial growth in space within a single generation that will return some $80 billion per year to the American economy. And that would be only the beginning. To turn our backs on that opportunity is to throw away our own future, and the future of our children.

To build the kind of space program that can help solve our economic and social problems on Earth, we must convince three sets of people to act with us: private industry, the government, and the environmentalists.

Of the three, perhaps it is most important to enlist the aid of the environmentalists. While most of them generally camp with the Luddites (and indeed, are probably the largest single group within that camp), the environmentalists' primary goal is to protect the natural ecology of Earth. They distrust the corporations and the government because centralized bureaucracies often put their own profits and procedures ahead of environmental protection: smokestacks ahead of trees, waterways projects ahead of the life of an endangered species.

The ultimate goal of the environmental movement and the new space movement is exactly the same: a clean, green planet Earth, a safe and fitting home not only for the human race, but for all the vast diversity of living forms that make up our world. The space program has already helped the environmental movement immeasurably. In the late 1960s, when our first Apollo explorers began to photograph planet Earth from the distance of the Moon, all of humankind saw our world as a beautiful jewel of life set in a dark void of barren cold. Men and women who had never before thought about their environment saw our home world as "the big blue marble" and realized that this island of life must be protected and preserved against pollution and ecological degradation.

Now space technology offers new opportunities for environmental protection of Earth. We can move much of our heavy industry off-planet. We can obtain our raw materials from the lifeless chunks of rock strewn through the Solar System, and thereby stop gutting our own world for metals, minerals, and timber. We can turn our home planet into a Class 1-A residential zone.

Today, most environmentalists are instinctively antispace. They see giant rockets bellowing through the Florida air as nothing but an elaborate form of pollution that threatens the ozone layer. Yet Cape Canaveral is not only the site of the Kennedy Space Center. It is also the home of one of the largest bird sanctuaries in the United States. The waterfowl of Florida get along quite nicely with the rockets of humankind. If only the humans could establish such an easy harmony among themselves!

We must win the environmentalists to the cause of space. Not all of them will be convinced that a vigorous space program will help lead to a cleaner, safer Earth. But those who will can help to show the rest of the population that a strong space program is a strong program for environmental protection, too.

The general Luddites' fear of corporate power makes them react against the space program, because there is no way an effective space program can be mounted without major contributions from the giant corporations. America's economic strength is based largely on the big corporations; they produce most of the goods and services we use in every part of our lives. They have the capital, the skilled personnel, the access to materials that are needed to build a new space program. We must avoid the either/or trap and realize that even though the corporations are necessary to the space program, it does not mean that the corporations will be the only ones to benefit from the program. We all will, if we run things properly.

Private companies are already investing in space. The West German firm, OTRAG, was launching its own test rockets from its base in Zaïre when the Angolan and East German–supported uprising in 1978 led to Zaïre's canceling OTRAG's lease. American entrepreneurs are quietly raising capital for private space ventures, and many major corporations have rented entire flights of the Space Shuttle so that they can test new industrial processes in orbit.

Before this decade is out, we may see industrial laboratories in orbit and privately owned "spaceports" being built in equatorial nations.

There is a danger that the giant corporations can become so powerful that they will be able to dictate the living conditions for us all. In this fear, the Luddites strike a bedrock stratum of truth. Anthropologist Marvin Harris, in his book *Cannibals and Kings* (Random House, 1977), warns against what he calls "the hydraulic trap." Harris points out that in ancient agricultural societies that depended on irrigation canals to water their crops, for example, the earliest civilizations of the Middle East and China, the central king who gained control of the canal network easily became an all-powerful tyrant. The power of the giant corporations may be just such a "hydraulic trap" for our civilization. The multinationals could acquire so much power that they become the *de facto* rulers of our lives. The question, then, is whether a strong space program — which necessarily requires heavy contributions from the major corporations — will lead to a future of more tyranny or more freedom.

I believe it will lead to more freedom for the individual, whether he or she goes into space personally or remains on Earth. By tapping new sources of wealth, by creating new opportunities for entire new industries, by widening the scope of human activity, space operations will lead to increased individual freedom. Tyranny flourishes where human activities are controlled, bound, circumscribed. Space is an ever-widening frontier that will constantly expand humanity's choices of action.

Indeed, the real "hydraulic trap" would be a world in which there is *no* space program at all, a world in which our growing numbers and dwindling resources force us to lay on more and more controls, tighter laws and regulations, more stringent penalties for breaking the rules.

In such a tightly controlled world, inevitably competition and initiative are throttled down and eventually stopped altogether. The "big fish" — be they giant corporations or national governments — would eventually swallow up the smaller private companies. So by keeping the space frontier open, we keep open the door for new entrepreneurs and the "little fish," as well. Tomorrow's IBM, Sony, TRW, and Texas Instruments will grow out of small high-technology firms that seize the challenge and the opportunity of space industry.

It is interesting to remember that the very idea of the corporation grew out of the challenges and opportunities to explore the New World. Financing expeditions to America or the Indies was a hugely risky business back in the fifteenth and sixteenth centuries. A wealthy man could be impoverished overnight if a ship foundered. So the concept of a corporation — a company in which the investors risked only what they actually invested, not their homes and private holdings — arose to become the backbone of the world's business.

Investing in space is also a huge risk, today. Perhaps new forms of business will be invented, as useful as the limited-liability corporation, to encourage private investment in the space frontier.

Although corporations and investors both large and small are studying the opportunities in space, today's space endeavors are still the exclusive preserve of national governments. In the US, it is the federal government that funds and directs all space operations. Washington's decisions about space throughout the 1970s were timid and short-sighted. The growing weight of opinion in America is to move the government toward a more vigorous and useful space program.

Even the General Accounting Office, a bastion of fiscal restraint, has reported that we are underspending in space. In a report released in 1979, the GAO stated that ". . . the U.S.'s conservative approach [in developing space for economic purposes] does not compare favorably to the specific long-term plans of the 11 members of the European Space Agency . . . nor to the plans of the Soviet Union and Japan." This prompted Thomas O. Paine, a former head of NASA and now president of Northrup Corporation, to comment, "When the government auditors say we've lost vision . . . we've lost vision."

Even the guardians of the purse strings are now convinced that we are hurting our economy by neglecting the opportunities in space.

It is vital, then, to build a coalition that embraces the diverse elements of space enthusiasts and environmentalists, giant corporations and kitchen-table entrepreneurs, government agencies and private citizens. As René Dubos urges, we must "Think globally; act locally." We must work as individuals to promote the expansion of space efforts, and to see that this expansion benefits the entire human race, not merely a chosen few. To accomplish this, we must *act*, whether the action is joining a local grassroots space organization or steering a multinational corporation.

The most important step that any individual can take at present is to help bring about a rapprochement between the environmentalists and the space enthusiasts. The movement to expand into space will take place with or without the environmentalists' help, but it will happen sooner, and be better directed, if the environmentalists do help. With their support, the goals of utilizing space technology to help solve the problems on Earth can be kept at the forefront of the new space program. If the environmentalists stay with the Luddite position that space is "bad" and must be opposed, it will force the space movement squarely into the embrace of the very entities the Luddites fear most: the big corporations and the Washington Establishment.

Working together, environmentalists and space enthusiasts can build a space program that will make the Earth greener and safer for us all. We can avoid the trap of swelling population and dwindling resources by expanding into a frontier as wide as the sky itself. The alternative is the politics of scarcity, the inevitable slide into decay and tyranny.

Other nations have faced this choice, ages ago. Nearly 500 years ago the Chinese developed deep-water navigation to the point where their ships ranged from Madagascar to the East Indies, exploring, trading, tapping the wealth of the Indian Ocean and the west Pacific. But abruptly the Ming Dynasty halted this flourishing commerce. No one really knows why, except that the central Chinese government decided to keep its people tightly bound to their own shores. Within a century China was being picked apart by Europeans, who found a backward, ignorant nation that was already fragmenting into petty principalities. What would have happened if a powerful, united, expanding Chinese empire had reached the divided, squabbling nations of Europe in A.D. 1450? Who can say how history would have gone then?

31

Provide for the Common Defense

> We hold that there is a real relationship be-
> tween a strong space program, the state of
> U.S. science and technology, and the eco-
> nomic and psychological health of the na-
> tion.
>
> —CHARLES SHEFFIELD

Dr. Sheffield is a space scientist and writer, a former president
of the American Astronautical Society, and a businessman
whose company, Earth Satellite Corporation, provides data
from Landsat-type spacecraft to commercial, industrial, and
other users.

In discussing space program objectives for the 1980s, Sheffield
has noted:

> In 1962, soon after the United States committed itself to the Apollo
> project, *Newsweek* cited a score of issues such as hunger, disease,
> and pollution that they gave more priority to than the "Buck Rogers
> stunt" of landing a man on the Moon. Now, nearly two decades later,
> the same publication notes that the post-Apollo decline in U.S. space
> funding has coincided with a drop in this nation's technological
> productivity. That drop has been matched by an increasing emo-
> tional and psychological malaise throughout the nation.

If by the term *the common defense* we mean the overall strength and well-being of the American people, then a vigorous space program can indeed provide for the common defense. We have seen in the preceding chapters exactly how space activities can help solve the economic and social problems we face, how a well-directed space program can counter the "emotional and psychological malaise" that grips many people.

How do we achieve these goals? Where do we start?

The Space Shuttle is the first step, as we have already seen. By reducing the cost of orbiting people and payloads, the Shuttle makes space industrialization possible. We must be prepared to face the emotional shock, however, of failures in the Shuttle testing program. No "hot" new aircraft has ever gone through its testing phase without at least one crash. If a Shuttle crashes, it will be much more than the lives of its crew and a billion dollars' worth of hardware lost. Confidence in the entire future of space operations will be shaken. Cries of Luddite despair will ring through the land, amplified by the media, to echo through the corridors of power in Washington. We must be prepared for setbacks, for failures, even for deaths. We must be prepared to turn back the defeatists and move onward.

With the Shuttle as an all-important first step, a vigorous space program for the 1980s should include:

1. Developing the Shuttle as swiftly as possible into a reliable and low-cost space transportation system that can carry personnel and equipment to low earth orbit as economically as possible.

2. Developing a Heavy Launch Vehicle, based on Shuttle technology, for orbiting even larger payloads even more economically.

3. Building a permanent space station in LEO, a Space Operations Center that will serve as a center for industrial experiments in materials processing, a research laboratory, a staging base for spacecraft refueling and maintenance, a medical labo-

ratory, a dormitory for work crews, and a center for large-scale construction projects. The LEO space station will gradually evolve and grow, much as a city does, to eventually become a tourist center and space hospital as well as a Space Operations Center.

4. Developing a "space tug," a vehicle that can carry payloads from the low orbits attainable by the Shuttle to higher, more useful orbits, and perhaps even as far as the Moon. Lunar landing vehicles could easily be developed from the basic "tug" hardware.

5. Increasing research on new space propulsion systems, solar energy, computers and automation, extraterrestrial mining, long-term closed-cycle life-support systems, and other technologies that contribute to space operations.

6. Establishing a manned station in geosynchronous orbit. The 24-hour "Clarke orbit" is the preferred site for communications, navigation, and meteorological satellites. Increasingly, as sensor technology improves, the planners will want to build larger, more sophisticated satellites and place them in geosynchronous earth orbit (GEO), where they can see half the world at once, rather than in lower orbits where the view is clearer but smaller.

This means that GEO will begin to fill up with satellites. A few permanent stations in GEO could consolidate many of these missions and allow continuous manned maintenance of the sensors. This could alleviate current worries that too many satellites will be competing for prime positions along the "Clarke orbit."

7. Planning and preliminary development of a permanent base on the Moon, where lunar mining and other research studies would be undertaken.

This is a powerful agenda for the 1980s. Although many politicians doubt that such a program could be initiated by Washington, there is little doubt among space enthusiasts that these objectives could be met, given the funding and national priority they deserve. Certainly, all these objectives cannot be met on a NASA budget that averages $5 billion annually. It is important to realize that, because of inflation, NASA's current budgets are *far* below the amounts received during the Apollo years. Five billion dollars in the mid-1960s equates to something like $14 billion in today's dollars.

While some of the more eager space enthusiasts feel that the seven-point program for the 1980s is too slow (they want to start building L-5 colonies *now*), it seems clear that the program detailed above could put us in a position to start building Solar Power Satellites and other truly large space artifacts in the 1990s. If there is a national consensus to pursue SPS, then the decade of the 1990s could see gigantic construction activities in space, and gigawatts of energy beamed to rectenna farms before the century closes.

A vigorous SPS program will lead almost necessarily to international cooperation in space. To begin with, there is no doubt that the SPS's will be financed by multinational corporations. The hundreds of billions of dollars of capital required means that only national governments or the largest of corporations will build SPS's. The US government might build the first prototype SPS on taxpayers' money, but after that it will be up to the corporations to build the rest — perhaps in cooperation with the government, as in the Comsat Corporation where private and public shareholders jointly own the company. Intelsat, the international comsat corporation, is jointly owned by 102 nations. These new semi-public, semi-private companies may well mark the path of the future financing of space. The markets for space-generated electricity will certainly be international. Developing nations will demand cheap power from SPS's perhaps even more loudly than industrialized nations that have access to fossil fuels and nuclear energy. The fears that SPS's might be turned into weapons will also force mandatory international cooperation, and inspection by international police teams. These teams will also be necessary to guard against terrorism, sabotage, or military action against SPS's. Thus the space technology of Solar Power Satellites may very likely force an internationalization of space operations, for reasons ranging from finance to security.

The 1990s may also see new technologies burgeoning, technologies that can dramatically change the way we move into space. Decades ago, Arthur C. Clarke calculated how much energy, in terms of kilowatts, it actually takes to lift an average-sized human being from the Earth and deposit him or her gently on the surface of the Moon. Then he asked his local electric utility how much that energy costs. He found that, in terms of the energy required, he should be able to travel to the Moon for a few hundred dollars. The fact that the Apollo project needed billions of dollars to do the job simply shows that rockets are not very efficient. Clarke is now telling everyone he meets about the possibilities of a space elevator, a true "skyhook," a tower 22,300 miles high in which you can ride to geosynchronous orbit in an electrically powered lift. Weird as the idea seems at first, calculations have proved that such a tower could be built, and it would be stable. Starting from a satellite construction platform in "Clarke orbit," the structure would be built downward until it reaches the Equator. When it is finished, we would have a literal fairytale beanstalk that climbs into the sky, and we could ride into orbit for pennies — because the electricity to run the elevator would be generated by solar cells along the tower's upper reaches.

The tower needs a structural material that is stronger than anything we possess today. The material must be about ten times stronger, weight for weight, than any structural material we now have. But as we saw in Chapter 21, *such materials could be developed in orbital factories.*

Ride an elevator 22,300 miles straight up? Go into space as easily as you ride up to a penthouse apartment? The prospect seems as wildly improbable as flying to the Moon seemed — once upon a time. Think of how the economics of space operations would change if such a space elevator became reality. All that we wish to do in space would cost a minuscule fraction of the costs we now foresee with rocket boosters and the Space Shuttle.

Even if the space elevator does not get built in the 1990s, other ideas percolating in today's research laboratories may reduce the costs of spacefaring drastically. It may be possible to boost rockets into orbit on the energy of very intense laser beams. This technology is in its infancy, but laser propulsion is being studied in several laboratories. Laser-boosted spacecraft could carry enormously larger payloads than today's chemically fueled rockets. And the powerful lasers that are being developed for military purposes may be the prototypes for this new and efficient propulsion system.

Once in the weightless and high-vacuum environment of orbital space, it could be possible to glide outward toward the Moon or planets on vast gossamer wings of solar sails. The minuscule force of sunlight, pushing on miles-wide sails, could provide the propulsive power for trips throughout much of the Solar System. Solar sail spacecraft would be slow, but very economical; the mass needed for propellants aboard a rocket craft could be given over to payload. Such sailcraft would make ideal freighters, carrying the heavy "pipeline" cargoes of ores and other natural resources from one human base to another.

These are a few of the possibilities of the 1990s, if we pursue a vigorous program of space development in this decade. There are other possibilities, however.

Even if the ambitious seven-point program for the 1980s is not carried out, both the US and USSR are pursuing military objectives in space. As we have seen, by the end of the 1980s both nations will have the capability of orbiting beam-weapon satellites that can destroy other spacecraft and shoot down rocket boosters. These weapons will require copious amounts of electrical power. Perhaps the military will develop their own version of the Solar Power Satellite, but it is more likely that they will develop power systems based on nuclear explosives and MHD (magnetohydrodynamic) generators.

These military programs are being carried forward today. By the 1990s there could be a silent, secret war going on a few hundred miles over our heads, totally unknown to the general public. The outcome of that war will determine who can carry through a nuclear strike on his enemies without suffering a catastrophic counter-strike. The winner of that silent war will dominate the Earth — as well as the entire Solar System's treasury of resources.

The alternative to a military battleground in space is a broadly based, highly visible, dynamic space program that is international in scale and global in its impact on human problems. The best way to provide for the common defense is to keep all the nations so busy getting rich that they no longer need to fight over territory or resources or prestige. If war is little more than organized theft, as Jacob Bronowski believed, then an open frontier of unending resources makes war nonsensical.

Can we internationalize space? The first tentative steps have already been taken. But they have created more controversy than cooperation, as we will see in the next chapter.

32

Promote the General Welfare

> I remember I was very sad for many days
> when I discovered that in the world there
> were poor people and rich people; and the
> strange thing is that the existence of the
> poor didn't cause me as much pain as the
> knowledge that at the same time there were
> people who were rich.
>
> —EVA PERON

Everything we have talked about in this book is based on the premise that the wealth we bring back from space will be shared in some manner with all the people of Earth.

But how? That is the central question of the new space age. In a world divided more bitterly every day into rich and poor, with the gap between the two widening instead of narrowing, with population rising along with tensions, hunger, wars, terrorism, how do we create an equitable sharing of the energy and resources to be found in space? The United Nations has attempted to solve that question by drafting a treaty that would govern all space activities. Already the treaty has created more rancor than harmony.

Officially titled "Agreement Governing the Activities of

States on the Moon and Other Celestial Bodies," unofficially it is called the Moon Treaty. Many Americans believe it may be the death knell for *any* effective American space program. They fear that the Moon Treaty will give the socialist nations an unbeatable advantage in space operations, and establish a new "OPEC-like monopoly" that will "control, regulate, and probably itself exploit outer space resources" while "erecting barriers to private American initiatives in space development." Supporters of the treaty say that this is an alarmist view, that the treaty will permit free enterprise in space (under international control), and that it is vitally necessary to establish a rule of law in space rather than first-come, first-served competition.

The major aim of the treaty, both sides agree, is to create a system for sharing the wealth that is found and developed in space. But opponents of the treaty feel that the system it would create will be so restrictive that no private entrepreneurs — not even the largest multinational corporations — will risk investing in space under such a deadweight of bureaucratic regulation.

If accepted by the US, the Moon Treaty will lay the clammy hand of delay on any corporation's plans for investing in space. And the delay could stretch out interminably. Space operations are risky enough without having an international bureaucracy waiting to take whatever profits you make and divide them up in some arbitrary manner.

Without private enterprise in space, the American space program will continue to be strictly a government affair, subject to the political fortunes of the White House and Capitol Hill, rather than the profit motive. Governments do not move briskly to exploit business opportunities. The benefits to be gained from lunar mining, space factories, Solar Power Satellites, and everything else we can do in space will be very slow in coming — if they come at all. Greedy and inefficient though they may be, capitalists develop wealth — it's the one thing they're good at. And the capitalist system has historically spread wealth among the general population much faster and more thoroughly than any socialist government has ever done.

It may be possible to amend the Moon Treaty to reflect US interests better. That is the view of many space law experts, although other experts point out that amendments or "understandings" attached by the US would have no international legal validity.

The Moon Treaty is an extension and elaboration of the basic principles of the 1967 Outer Space Treaty, which banned "weapons of mass destruction" in space and provided that the Moon will be a demilitarized area over which no nation may claim sovereignty. The first step toward the Moon Treaty was taken in 1970, when the Apollo program was at the height of its success. Argentina submitted a draft treaty covering the Moon and other celestial bodies to the legal subcommittee of the United Nations' Committee for the Peaceful Use of Outer Space (COPUOS). Argentina's draft was backed by the US, India, and Egypt. In 1971 the Soviet Union responded by submitting its own draft of a treaty. Before the year was out, the General Assembly recommended that COPUOS consider the Soviet draft.

For more than seven years the diplomats wrangled over the draft agreement. During this time, a group of Third World nations evolved the concept of "common heritage" regarding natural resources that are not within the territorial boundaries of any nation. It was the US delegation itself that placed a "common heritage" phrase in the first paragraph of Article XI of the Moon Treaty:

> The Moon and its natural resources are the common heritage of mankind which finds its expression in the provisions of this agreement . . .

Why did the American delegation insist on this wording? Partly for moralistic reasons. Partly as rhetoric intended to woo the Third World bloc at the UN. It was certainly a paraphrase of the words inscribed on the lunar landing module of Apollo 11, the *Eagle* that landed at Tranquility Base:

HERE MEN FROM THE PLANET EARTH
FIRST SET FOOT UPON THE MOON
JULY 1969, A.D.
WE CAME IN PEACE FOR ALL MANKIND

"For all mankind." That generous statement of faith now echoes throughout the corridors of the United Nations. For while the American delegates to the UN may insist that "common heritage" does not mean "common property," the Third World has a very different view. Ambassador M. C. W. Pinto of Sri Lanka said in 1978:

> The common heritage of mankind is the common property of mankind. The commonness of this "common heritage" is a commonness of ownership and benefit. The minerals are owned by your country and mine, and by all the rest as well . . . If you touch [them] in any way, you touch my property. If you take them away, you take away my property.

The "common heritage" concept was adopted in the mid-1970s by a political caucus of Third World nations that called itself the Group of 77. Now consisting of 132 nations, the Group of 77 published in the mid-70s a declaration calling for a "New International Economic Order" based on the rationale that "fundamental justice requires that those who receive the raw materials and natural resources which fuel and feed industrialized economies must be required to pay a significant share of their economic wealth in exchange for access to those resources."

Realize that the resources they are talking about do not lie within the territories of the nations who make this demand. They do not lie within the territory of any nation. This consortium of Third World nations, seeing the success of OPEC in escalating petroleum prices, decided to try the same tactics on resources *that they neither own nor have any hope of reaching.*

Their appeal to "fundamental justice" is an implication that the Third World nations are poor because they have been exploited by the industrialized nations. But a glance at history shows that these nations were poor five hundred years ago, when Europeans first colonized them. They were poor two thousand years ago, long before European discovery; read their own histories, or examine the lifestyles behind their mythologies.

In July 1979, ten years almost to the day after the first Apollo landing, the UN General Assembly passed the Moon Treaty unanimously. The American delegation voted "aye" along with everyone else. It was to go into effect as soon as it was ratified by five nations. Space advocates in the United States, led by the L-5 Society and *Omni* magazine, raised a storm of opposition, and it is extremely doubtful that the US will sign or ratify the treaty in its present form.

What is needed, these advocates insist, is a rewritten treaty that allows private companies to operate in space profitably. Otherwise, space will remain the domain of government bureaucracies (including NASA), and the economic benefits that can bring real wealth to the whole world — including the Third World nations — will never be realized. Government bureaucracies do not, and cannot, move quickly enough to produce commercial profits and economic benefits. Private enterprise does.

The USSR has not signed the Moon Treaty either, perhaps because compliance would oblige Russia to be much more open about its plans and activities in space, notifying the UN bureaucracy about future space operations and opening Soviet space facilities to international inspection. "Thank god for the L-5 Society," said one Russian negotiator, *sotto voce.*

Apparently the Soviets would sign the treaty if the US does, implying that the Russians would give up some measure of their traditional secrecy in exchange for stalling the expansion of US industrial enterprise into space.

Even the most adamant opponents of the treaty agree that the resources of space must be shared in some manner with the poor nations of the Third World. But they insist that a treaty that cripples free enterprise in space will strangle the best chance of rapidly developing those resources and bringing that wealth to Earth. We must share the wealth to be found in space, certainly. But we must also be free of bureaucratic controls so that we can reach the wealth and develop it. In this debate between "greedy" capitalists and "covetous" Third World politicians, some middle ground must be found.

The history of the UN Law of the Sea Treaty offers an instructive example of how *not* to handle space resources. Like outer space, the seabeds of the Earth's oceans are rich in natural resources. But the development of those resources was halted before it ever began because the nations could not agree on who owned what. For more than ten years the UN struggled to produce a definitive Law of the Sea Treaty, but that very same "common heritage" expression got in the way. In 1980 a Law of the Sea Treaty was passed by the UN. Among its provisions is the creation of an international "Enterprise" to govern and regulate seabed mining and determine how profits will be shared. This "Enterprise" is the equivalent to the Moon Treaty's "International Regime."

In 1980 the American government passed the Deep Seabed Hard Minerals Resources Act, a law that in effect gives government protection against legal suits against American companies that start to mine seabed minerals without UN approval. The US is only one of eight industrialized nations (the others are Great Britain, France, West Germany, Japan, Italy, Holland, and Belgium) that have passed similar legislation. Known as "the Like-Minded Group" they have essentially told their corporations to go ahead and mine the sea, no matter what the UN Treaty says.

Is this what we have to look forward to in space? Would it not be better to create a Moon Treaty that fairly and adequately protects the interests of both the industrialized nations and the Third World?

Apollo astronaut William Anders summarized the problem nicely:

> The Earth is not really the center of the universe. When you look at it from the Moon and it looks the size of your fist at the end of your arm, you don't see any international boundaries. If we on this grain of sand can't cooperate in space as mankind . . . and utilize this new medium for the benefit of us all, then we likely won't get together on anything and we'll bring about our own extinction.

33

Secure the Blessings of Liberty

> We set sail on this new sea because there is new knowledge to be gained and new rights to be won and they must be won and used for the progress of all people.
> —JOHN F. KENNEDY

When the Voyager 2 spacecraft sailed past Saturn with its dazzling, intricate rings and myriad moons, it marked the end of an era. It was, in Winston Churchill's words, "not . . . the beginning of the end. But . . . the end of the beginning."

Voyager's breathtaking reconnaissance of the outer reaches of our Solar System marks the end of Phase One of our space program. The first launch of the Space Shuttle, *Columbia,* marks the opening of Phase Two.

We have not finished the exploration of the Solar System, by any means. That task will take generations, centuries, to complete. But the explorations that began with Sputnik, Explorer, and Vanguard, and continued with Pioneer, Surveyor, Apollo, Venera, Viking, and Voyager, have taught us enough about the Solar System so that we can begin Phase Two with confidence.

Thanks to these explorations, we know that there are natural resources in superabundance throughout the Solar System: energy and raw materials for every industry on Earth, plus new industries that will be started in space itself. Now, as Phase Two starts, we will begin to use those resources to make life better on Earth.

Phase One of the space program was largely the domain of scientists, engineers, astronauts. Phase Two will need men and women of every profession and craft, from accountants to farmers, from construction workers to mechanics.

For what we will build in space is nothing less than a revolution.

We usually associate revolutions with violent upheavals, with war or the kind of civil strife that accompanied the first Industrial Revolution and the Luddites. But wars are not revolutionary. In my lifetime I have seen dozens of "revolutions" in nations scattered across Latin America, Asia, and Africa. In almost every case, after the coup d'état and the bloodletting, the form of government remains essentially the same; the only real change is that a new group of bullies takes over.

The real American Revolution did not take place in 1776, or at Valley Forge, Saratoga, Yorktown, or any of the other battlefields of the Revolutionary War. It took place in 1787, in my home town of Philadelphia.

Certainly the war fought by Washington's ragtag army was crucially important in driving British rule out of the thirteen American colonies. But the real revolution occurred when representatives of those states assembled in Philadelphia and hammered out a new Constitution, a new form of government, a truly new nation.

A revolution is a change in society, a move from something old to something new, a turn toward a different and better way of life. And that is what we will be doing as the new space program enters Phase Two.

In 1787, the Founding Fathers of this nation set a goal, an enduring task for all the generations to follow them. Rationalists that they were, they wrote this task down in lucid, concise language and placed it at the head of their new Constitution. This preamble states quite clearly that our ultimate goal is to "secure the Blessings of Liberty to ourselves and our Posterity."

Liberty. The freedom to live up to one's individual potential, to take one's rightful place in society, based on one's own abilities and desires.

In a circumscribed world of constantly expanding population and ever-dwindling resources, individual liberty is doomed. Even within the affluent United States, we have seen that when our economy falters, it is the poorest who suffer most, it is the blacks and Hispanics who are hit with the highest unemployment rate. Great social movements like civil rights and environmental protection grind to a standstill when the economy totters; we cannot *afford* to be fair with nonwhites, we are told; we cannot *afford* environmental protection.

That is why the new space program is important: because it will bring an economic bonanza not only to the United States, but to the world. Like it or not, the US is the leader of most of the world, and if our economy surges forward powerfully, the economies of Europe and all the globe will also advance.

These are the economic and practical benefits that Phase Two of the space program must bring to us. There are other, less tangible benefits as well. Freeman Dyson has written:

> The ultimate purpose of space travel is to bring to humanity, not only scientific discoveries and an occasional spectacular show on television, but a real expansion of our spirit.

We are stardust. Truly. The atoms that compose our bodies were created inside the ferocious maelstroms raging in the hearts of ancient stars. When those stars died they exploded and spewed their atoms into space, where they eventually became the infinitesimal building blocks of our Solar System, our Sun, our planet Earth, our very selves. Our urge to reach into space, to expand throughout the Solar System and beyond, to return to the stars, is as natural as the urge of a tree to seek sunlight. Dust to dust: Stardust we are, and to the stars we seek to return, knowing instinctively that to be bound to the surface of one planet is to be chained to limits that will ultimately destroy us.

Consider the old maxim: It is better to light one candle than curse the darkness. Think of the subtlety of that statement; the assumptions it contains. Action toward a goal is more desirable than rhetoric. More important, the action to be taken is a *technological* action. It is so simple and logical, so natural: Light a candle. We solve our problems with technology, our ability to use fire, to make candles, to create light where there was nothing but darkness.

Phase Two of the space program can bring us the lunar mines, the prospector ships sailing out to the asteroids, orbital factories, L-5 colonies, the expansion of the human spirit — all that transforms us into a truly spacefaring people. It can also bring the laser-armed satellites, the silent war on orbit, the struggle to dominate the Earth with weapons in space.

How do we direct our efforts to produce a space program that brings peace and abundance to Earth?

Look back a moment: Two decades ago, 1961. President John Kennedy came into office to find the Russians far ahead of us in space successes: the first artificial satellite, the first flights to the Moon, the first man in space, the first interplanetary probes. The Kennedy team decided to leapfrog the Soviets and aim for a manned landing on the Moon. Project Apollo was born.

In the sense of infantry tactics, this was "taking the high ground." The technical challenges of putting astronauts on the Moon and returning them safely to Earth forced us to create a technology that, once forged, could take us anywhere in the Solar System we wished to travel. The challenge of Apollo focused our space efforts and allowed us to leap into the forefront of space activities.

Now it is time for a new challenge, a new goal that will focus all our abilities and direct the development of our space technology toward an objective that is not only worthwhile in itself, but which will spin off many additional benefits.

I suggest that the President of the United States set this objective for our space program: that we build a prototype Solar Power Satellite system before the year 2000. Technically, this is a challenge that we can meet. The basic technologies that make up the SPS system are known; the engineering can be done. Politically, this is a goal that transcends normal party politics, because its culmination is nearly two decades away. No President and very few members of the Congress can hope to be still in office when the first SPS goes into operation.

The spinoffs from such a project would be monumental. First, the prototype SPS would be used to determine exactly what are the environmental effects of such a system. The rectenna farm would be located in a remote area, such as the White Sands Proving Grounds, in New Mexico, so that no possible danger to people would arise. Second, to build and assemble an SPS system would require boosters, construction equipment, a vast team of people and machines that could then be used to build other large structures in orbit: factories, laboratories, zero-gee hospitals, even permanent habitats.

The cost would be large, perhaps as much as one or two percent of the total federal budget. Not as much as we spend on cigarettes or alcohol, but large. The payoff would be even bigger.

One prototype SPS in orbit and operating by the year 2000 A.D. That is a goal worth working toward.

*

We have journeyed through a long march, you and I, traveling together from the earliest beginnings of our work in space to a future where space operations can help mightily to solve Earth's problems of hunger, ignorance, poverty, and war. I have no way of knowing if I have convinced you of my case. History will give us both the answer to that, but only at the maddeningly slow one-second-per-second pace at which history always moves.

How will we shape our destiny? Will we take the bold, active role and — like Beethoven — seize fate by the throat? Will we use the best that is in us to create a future of abundance and freedom, of triumph over our most ancient and remorseless enemies? Or will we passively accept hunger, poverty, ignorance, and death as the inescapable victors over the human spirit?

I believe that there is not a human baby born but can move heaven and earth with the power of its goodness and intellect. But I am a writer, a dreamer; and practical people know that such men and women are inclined to foolish visions. Yet out of such foolishness comes all the greatness of our kind. As I write these words I am flying six miles above the heartland of America, moving at nearly the speed of sound, buoyed by the dreams, the work, the devotion of two brothers who repaired bicycles in the unlikely town of Dayton, Ohio.

It is a strange and marvelous world. We are capable of great and marvelous achievements. What is past is mere prologue. The best is yet to come, if we dare to reach for it.

There are great deeds to be done, whole planets to explore, whole worlds to be built. The gestation of the human race is about to end and we will be born into a wide and starry universe. Even if Americans do not take part in this cosmic awakening, other people will: the Europeans, the Japanese, the Chinese, the Russians.

The Russians.

Late in 1980, the Soviet Union launched a new type of manned spacecraft into orbit, where it docked with the Salyut 6 space station. This new craft, called the Soyuz T-3, was the first Soviet spacecraft to carry three cosmonauts since the ill-fated Soyuz 11, in which three cosmonauts died during re-entry, in 1971. The Soyuz T-3 is a sure indication that the Russians are moving ahead with their orbital operations.

As you read these lines, it is likely that the USSR will be constructing a permanent 12-man space station in low earth orbit, to follow up the success of its outstanding Salyut 6. The Soviets are also testing a delta-winged shuttle-type aerospace plane, smaller than the US's Space Shuttle, but reusable and reliable.

The future belongs to those who build it. Neither Americans nor any other people have an assured birthright to tomorrow. It would be the cruelest irony in human history if the Russians, of all the tyrannized people on the face of this Earth, find prosperity and freedom in space, while we in "the land of the free" sink deeper and deeper into poverty, despotism, and despair.

Index